This book represents a major contribution to game theory. It offers a new conception of equilibrium in games: strategic equilibrium. This new conception arises from a study of expected utility decision principles, which must be revised to take account of the evidence a choice provides concerning its outcome. The argument for the new principles distinguishes reasons for action from incentives, and draws on contemporary analyses of counterfactual conditionals. The book also includes a procedure for identifying strategic equilibria in ideal normal-form games.

In synthesizing decision theory and game theory in a powerful new way this book will be of particular interest to all philosophers concerned with decision theory and game theory as well as economists and other social scientists.

T0275635

Equilibrium and Rationality

Cambridge Studies in Probability, Induction, and Decision Theory

Equilibrium and Rationality

Game Theory Revised by Decision Rules

Paul Weirich
University of Missouri – Columbia

CAMBRIDGE
UNIVERSITY PRESS

CAMBRIDGE UNIVERSITY PRESS
Cambridge, New York, Melbourne, Madrid, Cape Town, Singapore, São Paulo

Cambridge University Press
The Edinburgh Building, Cambridge CB2 8RU, UK

Published in the United States of America by Cambridge University Press, New York

www.cambridge.org
Information on this title: www.cambridge.org/9780521593526

First published 1998
This digitally printed version 2007

A catalogue record for this publication is available from the British Library

Library of Congress Cataloguing in Publication data
Weirich, Paul, 1946–
Equilibrium and rationality : game theory revised by decision
rules / Paul Weirich.
p.   cm. – (Cambridge studies in probability, induction, and
decision theory)
Includes bibliographical references and index.
ISBN 0-521-59352-2 (hb)
1. Game theory.   2. Statistical decision.   I. Title.   II. Series.
QA269.W46   1998
519.3 – dc21   97-16555
                                                              CIP
ISBN 978-0-521-59352-6 hardback
ISBN 978-0-521-03802-7 paperback

For my parents

Contents

Preface

Game theory is a rich field for the philosopher. It formulates principles of rational action and so makes immediate contributions to the branch of logic that studies practical reasoning. Its principles of strategic reasoning also provide the undergirding for the social contract theory and the study of social and linguistic convention. Revisions of central game-theoretic tenets reverberate throughout philosophy.

This book makes some assumptions about rationality and equilibrium in games and traces out their consequences. These are the main assumptions: (1) There are no dilemmas of rationality; an agent has a rational choice in every choice situation. (2) Every ideal game has a solution, that is, a profile of jointly rational strategies. (3) A rational choice is self-supporting. (4) An equilibrium is a profile of strategies that are jointly self-supporting. These assumptions suggest replacing Nash equilibrium, which fails to exist in some ideal games, with a new type of equilibrium called *strategic equilibrium*. Strategic equilibrium is introduced via a study of self-support. Taking an equilibrium to be a profile of jointly self-supporting strategies, I use a new account of self-support to obtain the new type of equilibrium. I show that a strategic equilibrium exists in every ideal normal-form game meeting certain mild restrictions and provide procedures for finding strategic equilibria in those games.

This project arose from reflections on cooperative games, where individuals may form coalitions that adopt joint strategies and otherwise act as agents. I took being an equilibrium as a necessary condition for being a solution to a cooperative game, and this led to a problem. The most widely accepted account of equilibrium for cooperative games takes a game's outcome to be an equilibrium just in case it is in the game's "core," and so is an outcome such that no coalition can increase its payoff by unilaterally changing its strategy. But many cooperative games have empty cores, and so no solutions according to that account of equilibrium. To address this problem, Aumann and Maschler (1964) advance a type of equilibrium known as a "bargaining point" that may exist outside a game's core. Their main idea is to take some incentives to be defeated by objections to their pursuit. A similar idea also leads Bernheim et al. (1987) to introduce for games with certain cooperative elements a type of equilibrium called "coalition-proof Nash equilibrium."

ix

To investigate defeated incentives in cooperative games, I applied the decision principle of weak ratifiability introduced in Weirich (1988) to cooperative games with an infinite number of salient outcomes, for instance, an infinite number of divisions of some resource. The conclusions about solutions that emerged, however, are clearest for noncooperative games. Complications concerning the treatment of coalitions as agents obscure them in cooperative games. To throw these conclusions into high relief, this book focuses on noncooperative games. It uses the general idea of defeaters for incentives to define equilibrium for noncooperative games. In certain cases where an agent's pursuing an incentive to switch strategy generates incentives for other agents to switch strategy, I say that the agent's incentive is not a sufficient reason to switch strategy – it is defeated. Thus a strategy need not be incentive-proof to be self-supporting and part of an equilibrium. Although I define equilibrium in a detailed way for noncooperative games only, I envisage extending the definition to cooperative games as well.

My goal is to use principles of rational decision, in particular, principles of self-support such as the principle of weak ratification, to explicate equilibrium, taken as a necessary condition for a solution. I originally planned to derive the realization of a Nash equilibrium from principles of self-support, using idealizations about the rationality and common knowledge of the players. But that project required many restrictions on the type of game treated. To gain generality, I introduce strategic equilibrium, a new, weaker type of equilibrium. And to simplify, I focus on its existence rather than its realization and adopt an idealization about the players' foreknowledge of the game's outcome rather than derive that idealization from more basic idealizations about the players' common knowledge.

The book has three main parts. First, it presents the problem that motivates its proposals – the absence of Nash equilibria in some ideal games – and explains the nature of solutions and equilibria in games. Second, it advances a new decision principle of self-support, applies the principle to noncooperative games to obtain a new type of equilibrium, and shows that the new type of equilibrium exists in every ideal normal-form game meeting certain restrictions. Third, it formulates a procedure for finding equilibria, applies it to examples, and compares the new equilibrium standard for solutions to other familiar standards.

Chapter 1 presents the problem of missing Nash equilibria in a preliminary way and explains our conception of games and solutions. Chapter 2 discusses the idealizations we adopt for games. Chapter 3 shows why the absence of Nash equilibrium in ideal games is damaging for that conception of equilibrium and in general lays out the reasons for a revised conception of equilibrium. In particular, it examines problems with the allied standard of incentive-proofness for solutions. This standard says that a solution must be such that no agent has an incentive to switch from his part in the solution given the parts of other agents. In a noncooperative game in which agents choose independently, every switch in strategy is unilateral. Thus the standard requires that no agent have an incentive to switch from a solution. The standard of incentive-proofness for solutions supports Nash equilibrium as a necessary condition for solutions in ideal noncooperative games. Because of the problems with the standard, we reject it and the companion standard of Nash equilibrium.

Chapter 4 argues for the decision principle of self-support used to flesh out the new equilibrium standard for solutions and rejects competing principles of dominance, ratification, and expected utility maximization. Its arguments undercut decision-theoretic support for Nash equilibria and open the door to a new, less restrictive type of equilibrium. Chapter 5 presents the new type of equilibrium and demonstrates its existence in all ideal normal-form games meeting certain simplifying assumptions.

Chapter 6 formulates a procedure for finding the new type of equilibrium. Chapter 7 applies the new equilibrium standard to games of special interest. Chapter 8, the final chapter, compares the new equilibrium standard with other standards, including the standard of Nash equilibrium and the standard of nondomination, the two most widely endorsed standards for solutions.

The book's main foundational point is the argument for the new principle of self-support, and its main technical points are the demonstrations of the existence of the new type of equilibrium in ideal normal-form games and of the adequacy of the search procedure for equilibria of the new type. Other important auxiliary claims are the following: (1) Decision principles should be adjusted to recognize the futility of pursuing certain types of incentive. It is a mistake to require that all incentives be pursued. The principle of incentive-proofness makes sense only where pursuit of incentives is not futile. (2) An agent's choice should use his knowledge of his pursuit of incentives just as it should use his knowledge of other agents' pursuit of incentives. In particular, it should use his knowledge of his responses to other agents, just as it should use his knowledge of other agents' responses to him. (3) Equilibrium in a noncooperative game may be defined in a way that makes equilibria depend upon features of a game not represented by its payoff matrix. For example, equilibria may depend upon the way ties between strategies are broken.

The book's main limitations are the following: (1) Only ideal normal-form games, in which agents are prescient concerning other agents' choices, are examined in depth. (2) The idealization of prescience is not derived from more basic idealizations concerning common knowledge and the like. (3) Only the existence, not the realization, of equilibria is demonstrated.

Although technical material is self-contained, some general background reading will help those unacquainted with decision theory and game theory.

See, for example, Luce and Raiffa (1957), Jeffrey (1983), and Resnik (1987).

I gratefully acknowledge support during 1994–5 by the University of Missouri Research Board and the University of Missouri-Columbia Research Council. For helpful comments, I am indebted to Peter Markie, Wlodek Rabinowicz, Howard Sobel, Xinghe Wang, and two anonymous readers. Louise Calabro and Susan Thornton provided invaluable editorial assistance.

1

Games and Solutions

Game theory and decision theory have a symbiotic relationship. Game theory motivates revisions in principles of rational choice. And decision theory motivates revisions in accounts of solutions to games. My project is to revise game theory in light of reflections on rational choice. I provide a new account of equilibrium in games – equilibrium among strategic reasoners – motivated by equilibrium's connection with solutions and rational choice.

To explain equilibria and solutions, let us examine a game of Hide and Seek with a time limit. The Seeker has to find the Hider before time expires; otherwise the Hider wins. The Hider can conceal herself on either the first or the second floor of a house. The second floor is smaller than the first so that the Seeker can search all of it before time expires, whereas he can search only half the first floor in the time available. The second floor has windows from which, looking down through skylights, parts of the first floor can be seen. As a result, half the first floor can be searched while searching the second floor. The players know about the windows but do not know which parts of the first floor are visible from the second floor.

Shuttling between floors would be time wasting for the Seeker. His only effective strategies are (1) searching the first floor and (2) searching the second floor. The Hider also has to choose between the first and second floors. Suppose that the players know that the chances of the Hider's being found are as follows for each of their strategy combinations: (a) 100% if the Hider is on the second floor and the Seeker searches there, (b) 0% if the Hider is on the second floor and the Seeker searches the first floor, (c) 50% if the Hider is on the first floor and the Seeker searches there, and (d) 50% if the Hider is on the first floor and the Seeker searches the second floor. Figure 1.1 summarizes these probabilistic outcomes of their strategy combinations. What should the players do given that each is psychologically astute and able to anticipate the other's strategy?

Plainly, each player should decide in light of the other's choice. This rules out some strategy combinations. If the Hider goes to the second floor, then, being an accurate predictor, the Seeker looks there. The Hider has an incentive to switch floors; her strategy is not self-supporting. Only if the Hider goes to the first floor and the Seeker searches the second floor are their strategies jointly self-supporting. Such an outcome is an equilibrium since

Hider

		Floor 1	Floor 2
Seeker	Floor 1	50% discovery chance	0% discovery chance
	Floor 2	50% discovery chance	100% discovery chance

Figure 1.1   Hide and Seek.

Mismatcher

		Heads	Tails
Matcher	Heads	Matcher wins	Mismatcher wins
	Tails	Mismatcher wins	Matcher wins

Figure 1.2   Matching Pennies.

it balances the strategic considerations of the players. Joint self-support is a prerequisite of joint rationality, which is the hallmark of a solution. So the game's unique equilibrium – hiding on the first floor and seeking on the second floor – is also its unique solution. Here a solution is understood in a subjective sense that makes it depend on the rationality of strategies rather than their successfulness. The game has no solution in an objective sense according to which a solution gives each player success. Whatever the players do, one wins and one loses.

1.1. MISSING EQUILIBRIA

Now consider another game, Matching Pennies. This game has two players, each of whom has a penny he can display with either Heads or Tails showing. The players display their coins simultaneously. One wins if the coins match; the other wins if they do not match. Figure 1.2 depicts their situation. Suppose that the players cannot randomize their choices, and each will anticipate the other's move. What should they do? This game has no strategy combination in which each strategy is a best reply to the other. If both players display Heads, for example, the mismatcher does better with Tails. What are the consequences of such games for an account of solutions and equilibria?

The outcome of a game is a strategy for each player. A (subjective) solution to a game is an outcome in which the players are jointly rational. Does every game have a solution? I hold that it is always possible to choose

2

rationally. I deny the existence of dilemmas of rationality in which all choices are irrational. I do not argue for this view but take it as a starting point. This view motivates the related but independent view that every ideal game has a solution. I argue that since the players in a game do not face dilemmas of rationality, their joint rationality is possible in ideal games; that is, a solution exists. The restriction to ideal games is essential since idealizations concerning the rationality and knowledge of the agents are needed to remove obstacles to the existence of a solution. The possibility of individual rationality for each agent is not sufficient by itself to ground the possibility of joint rationality for all agents. But, I contend, individual rationality (Chapter 4's topic) and appropriate idealizations (Chapter 2's topic) together ensure the existence of a solution.

An equilibrium of a game is an outcome in which the strategies adopted by the players are jointly self-supporting. Since a rational strategy is self-supporting, a solution of a game has strategies that are jointly self-supporting. In other words, a solution is an equilibrium. Since every ideal game has a solution, it also has an equilibrium. This conclusion forces a revision of views about equilibrium.

The common view takes equilibrium to be Nash equilibrium. This view interprets self-support in ideal games, in which players are aware of what others do, in terms of best replies. A *Nash equilibrium* of an ideal game is an outcome in which each player adopts a best reply to the others. There are two ways to interpret the standard of Nash equilibrium. The *objective* interpretation states the standard in terms of payoff increases: An outcome is a Nash equilibrium if and only if no strategy change by a single agent produces a payoff increase for the agent. The *subjective* interpretation states the standard in terms of preferences of the agents: An outcome is a Nash equilibrium if and only if given the outcome, no agent prefers an outcome that he can reach by changing strategy unilaterally. A subjective Nash equilibrium is an outcome that is *incentive-proof*, or free of (subjective) incentives to switch strategy. The objective interpretation is the canonical one. But the standard is usually advanced for ideal games, where agents are rational and informed about the game. The idealizations make payoff increases and incentives coincide so that the objective and subjective interpretations agree. In ideal games the objective Nash equilibria are exactly the subjective Nash equilibria. I generally adopt the subjective interpretation of the standard, since it makes more explicit the connection between Nash equilibrium and rationality requirements for agents. But I also generally limit myself to ideal games, in which the two interpretations of the standard agree. Consequently, the distinction between the two interpretations of the standard is not critical for my purposes.

Some ideal games, such as the version of Matching Pennies discussed, lack Nash equilibria. The subjective interpretation of the standard of Nash equilibrium insists on the absence of agents' incentives to switch strategy. But the standard cannot be met in all games. This creates a problem given the foregoing views about solutions and equilibria. Here is the problem in a nutshell. Every ideal game has a solution. Moreover, only equilibria are solutions since self-support is necessary for rationality. Consequently, every ideal game has an equilibrium. But, uncontroversially, there are games without Nash equilibria and, I argue, also ideal games without Nash equilibria. Therefore the common view of equilibrium must be revised to provide for the existence in ideal games of equilibria taken as combinations of strategies that are jointly self-supporting. We need an account of equilibrium weaker than Nash equilibrium that allows for the existence of equilibria in ideal games. The full argument for revising contemporary views about equilibrium takes us to the end of Chapter 3. Chapters 4–8 work out the revision.

The addition of randomized or *mixed* strategies guarantees the existence of a Nash equilibrium in a *finite* game (a game with a finite number of agents and of unrandomized or *pure* strategies for each agent). At least it does so under the usual assumption that the value of a mixed strategy is the probability-weighted average of the values of component pure strategies. Nash's famous theorem (1950) establishes this. The theorem stands even if an agent's mixed strategies are taken, as in Harsanyi (1973), as probability mixtures only with respect to other agents' probability assignments. Nonetheless, the problem of missing Nash equilibria remains for games without mixed strategies (because of the unavailability of randomization or the predictive powers of agents) and for games with an infinite number of pure strategies for some agent. A general theory of equilibrium must address such games.

Some theorists quickly dismiss the problem. According to the common view that a Nash equilibrium is a solution, a game with no Nash equilibrium is simply a game with no solution. However, this response runs contrary to the intuition that every ideal game has a solution (putting aside special cases with an infinite number of agents) and wreaks havoc with the theory of rationality, as we shall see. We need to explore alternative, less devastating responses to the problem. We need to reexamine rationality requirements for strategies, and equilibrium standards for solutions.

Equilibria as generally conceived are absent in ideal games because of the stringent type of self-support used to define equilibria. Commonly a strategy is said to be self-supporting if and only if it maximizes expected utility on the assumption that it is adopted, in other words, if and only if it is incentive-proof, or ratifiable. This standard of self-support is too high. It cannot be met in all decision problems. Chapter 4 provides examples. Our approach to the problem of missing equilibria is to lower the standard of self-support. We obtain a weaker, more easily attained equilibrium standard by using a less demanding type of self-support to fill out the definition of equilibrium.

I modify the definition of an equilibrium, a profile of jointly self-supporting strategies, by taking *self-support* as the absence of a sufficient reason to switch strategy, but not necessarily the absence of an incentive to switch strategy. As Chapter 4 argues, not every incentive to switch strategy is a sufficient reason to switch. An incentive may be insufficient, for instance, if it is undermined by the supposition that it is pursued, in particular, if given its pursuit the other agents' response undermines the incentive. Our new type of equilibrium, motivated by decision principles concerning self-support, is attainable in ideal games.

The new account of self-support considers paths of incentives to switch options, say an incentive to switch from A to B, from B to C, and so on. A path of pursued incentives terminates if it has a last member and no extension. To be sufficient in the games treated, an incentive must go somewhere; it must start a terminating path of pursued incentives. As Chapter 4 argues, in these games a self-supporting strategy is one that does not start a terminating path of pursued incentives. Chapter 5 applies this view to equilibria – they must be composed of strategies that are jointly self-supporting – and shows that all games of a certain type have an equilibrium. Chapter 6 shows that in some of those games a profile is an equilibrium if and only if no agent pursues an incentive away from it and uses this feature of equilibria to identify them. Chapter 7 explains that an ideal version of Matching Pennies has an equilibrium dependent on the players' pattern of pursuit of incentives. If, for instance, the mismatcher abandons endless pursuit of incentives at the point where both players show Heads, then that point is an equilibrium.

My main concern is the existence of equilibria. Showing that equilibria exist is of course different from showing that they are realized. Showing that equilibria exist in a game requires showing the possibility of joint self-support. It is another matter to show that the possibility is realized, and that the outcome of the game is an equilibrium. I put aside the realization of equilibria. I do not consider whether given appropriate idealizations rational agents realize an equilibrium. Nor do I consider whether in games with multiple equilibria, rational agents do their parts in a particular equilibrium. I offer no principles of equilibrium selection. The problem of equilibrium selection is challenging, and is exacerbated by our new equilibrium standard since it provides equilibria in addition to Nash equilibria from which to choose. Still, I consider only whether given appropriate idealizations equilibria exist. Rather than use decision principles to derive the realization of an equilibrium, I use them to establish the existence of an equilibrium. Instead of deriving the realization of an equilibrium from compliance with decision principles, I reduce being an equilibrium to compliance with certain decision principles. Then I show the possibility of attaining the equilibrium standard by demonstrating the possibility of compliance with the decision principles. Explaining the realization of equilibria is an important project, but a project for another occasion.

One may wonder whether the issues of existence and realization of equilibrium can be separated as I propose. Doesn't support for a definition of equilibrium require showing not only that equilibria exist in ideal games but also that an equilibrium is realized? If certain profiles are jointly self-supporting in ideal games, then one of them must be realized. The realization of other profiles is incompatible with the agents' rationality and knowledge. To verify the definition of equilibrium, shouldn't we explain how principles of individual rationality lead each agent to his part in one of the equilibria? This verification is indeed desirable. But it requires a much broader account of rationality than an account of self-support, and a much broader account of agents' knowledge in ideal games than we provide later. To make our project manageable, we have to forgo this type of verification. The case we make for the revised definition of equilibrium is nonetheless strong.

1.2. NORMAL-FORM GAMES

The equilibrium standard for solutions applies to all games since it stems from the general rationality requirement that an agent's choice be self-supporting. However, our investigation of equilibrium is limited to a certain class of games – normal-form, noncooperative, single-stage games of strategy with a finite number of agents. The terminology used to define this class of games is from standard works in game theory; it is reviewed to inform those new to the field and to make precise the interpretation adopted here. These games clearly bring out a problem with interpreting self-support as incentive-proofness and hence equilibrium as Nash equilibrium.

A *game* is a decision situation involving two or more agents where each agent's outcome depends on the actions of other agents as well as his own action. This definition covers familiar games such as chess, but also competition between firms, coordination between drivers in traffic, bargaining over a division of resources, and many other types of personal interaction. The definition may be extended to include degenerate cases where there is just one agent, or the outcome for an agent depends only on his own action. Then any decision situation for any number of agents counts as a game. But the central cases, and the ones that interest us, involve multiple agents with the outcome for each agent dependent on the actions of all.

We call the options available to an agent in a game *strategies*. The strategies may be simple actions or complex conditional actions replete with contingency plans and the like. We use the term *strategies* because the reasons for an agent's action in a game are often strategic in a broad sense, looking ahead to the responses of other agents. We call an assignment of strategies to agents, one strategy for each agent, a strategy *profile*. A strategy profile determines the outcomes or *payoffs* of the game for the agents. Since a strategy profile determines the outcome, it may substitute for a possible outcome.

We focus on games in which rational decision relies on principles of strategic reasoning. We call these *games of strategy*. We put aside other games, for example, repeated games, in which rational decision requires learning about opponents' future choices from their past choices.

As in nondegenerate games of any sort, in a game of strategy the outcome of one agent's strategy is dependent on other agents' strategies. A situation in which two agents simultaneously choose from a dinner menu, for example, does not constitute a game of strategy. Also, and more important, in a game of strategy an agent's choice affects his preferences among strategies, or his incentives to adopt strategies. Given one choice an agent may have an incentive to adopt a certain strategy, but given another choice that incentive may disappear. An agent's incentives to adopt strategies are not constant given his possible choices, in contrast with most decision situations. In games of strategy rational agents use strategic reasoning to take account of the anticipated effect of a choice on incentives to adopt strategies. To explain these games, let us consider the characteristics of strategic reasoning.

In many games of strategy incentives are inconstant with respect to strategies because the agents can anticipate, at least to some degree, the responses of other agents to their strategies. An agent's adoption of a strategy provides evidence about other agents' strategies. Rational agents use strategic reasoning to make their choices because it takes account of anticipated responses.

Strategic reasoning involving anticipation of responses is most clearly exemplified in games where an agent makes a move observed by other agents, and then the other agents make moves observed by the agent, and so on. For instance, in Tic-Tac-Toe, if the agent who plays first, X, takes the middle, then the agent who plays second, O, takes a corner, reasoning that if she takes any other spot, X will respond in a way that brings him victory as in Figure 1.3, where X's first move is X1, his second move is X2, and so on, and similarly for O. O uses her strategically inferred information about X's responses to her strategies to guide her selection of a move.

Some games do not have a sequence of publicly observed moves. In *single-stage games* all agents choose during a single period using initial information about the game, and their choices are not made public until the period ends. Games where choices are made simultaneously are single-stage. But simultaneous choice is not essential. The characteristic feature of single-stage games is that no agent can observe the choices of other agents before choosing. Information about their choices must come, directly or

7

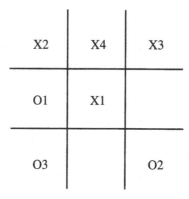

Figure 1.3   Tic-Tac-Toe.

by inference, from initial information about the game. I treat single-stage games of strategy.

In single-stage games of strategy, strategic reasoning may involve an agent's anticipation of other agents' choices, and his anticipation of their choices may depend on his information about their abilities to anticipate his own choice. An agent may know that other agents anticipate a certain choice by him and that they make their choices according to what they anticipate. For instance, in Scissors-Paper-Stone, an agent may know that his opponent anticipates Paper from him and that she will play Scissors as a result. The agent may then surprise his opponent by playing Stone.

Strategic reasoning in single-stage games of strategy is analogous to strategic reasoning in multistage games, except that instead of considering the response of other agents to a move, an agent considers the evidence a strategy provides for other agents' strategies. Strategic reasoning plays a role via foresight of the incentives that obtain if a choice is made. An agent has foresight of those incentives either because of anticipation of the other agents' response to his choice or because of anticipation of changes in the circumstances for incentives given his choice. An agent's incentives are relative to his choice, and strategic reasoning uses knowledge of this relativity to make a choice. Since the analogy between strategic reasoning in single-stage and multistage games is strong, we use the terminology of strategic reasoning in multistage games to describe strategic reasoning in single-stage games. We speak of the "response" of other agents to an agent's strategy in a single-stage game if his strategy is conclusive evidence for certain strategies by other agents, even though, strictly speaking, because his strategy has no causal influence on their strategies, they do not respond to his strategy.

In recent literature Nash equilibrium is generally taken as the appropriate conception of equilibrium only for single-stage games, such as Matching Pennies, as opposed to multistage games, such as chess. Multistage games raise issues of strategy ignored by Nash equilibrium, as Selten (1965) observes. Although I believe that our new type of equilibrium applies to multistage games, I do not work out the details. Its main advantage over Nash equilibrium in multistage games is that it draws on more features of a game than does Nash equilibrium. Nash equilibrium takes into account only the features represented by a game's *payoff matrix*, that is, a table giving the payoff for each agent of each strategy profile. The new type of equilibrium takes account of other strategic considerations ignored by Nash equilibrium.

I focus on a particular type of single-stage game of strategy – *noncooperative* games. Noncooperative games are sometimes characterized as games without opportunities for binding agreements. It is better to take them as games where the agents are not coalitions or are not treated as coalitions. Binding agreements are just a means of forming coalitions that act as agents. The inability of agents to form coalitions and act jointly may arise because they are unable to communicate, because their interests conflict, or because their agreements cannot be made binding.

It is also better to apply the distinction between the cooperative and the noncooperative to representations of games rather than to games themselves. A representation of a game standardly identifies, among other things, the agents and their strategies. In a noncooperative representation of a game, the agents are individuals, or groups treated as individuals. In a cooperative representation of a game, the agents are coalitions, including single-membered coalitions, of individuals. The coalitions are treated as coalitions, not as individuals; such treatment entails acknowledging that they may not form, and so assigns them the strategies of formation and nonformation. A single game may receive both a cooperative and a noncooperative representation. A game with a cooperative representation may be represented as a noncooperative game by reducing the strategies of coalitions to combinations of strategies of individuals. Generally, the reduction requires treating a game with a single-stage, cooperative representation as a multistage, noncooperative game in which individuals make offers of collaboration, counteroffers, and the like, before they realize a coalition structure and joint strategies for its coalitions. The strategies assigned to the members of a coalition by a profile in the game's noncooperative representation constitute a strategy for the coalition in the game's cooperative representation.

We treat noncooperative games because Nash equilibrium is generally proposed as the appropriate conception of equilibrium for them and because noncooperative games form the fundamental type of game. Cooperative games are in principle reducible to noncooperative games. Although

I investigate equilibrium in noncooperative games only, my definition of equilibrium is intended for cooperative games as well. Its extension to cooperative games faces two issues, however. First, one must fully work out a method of treating a coalition as an agent so that self-support is precisely defined for a coalition's strategies. Second, one must specify the idealizations that are appropriate for cooperative games in order to verify that equilibria exist in all ideal cooperative games. I leave these two issues open but assume they can be settled favorably.[1]

In noncooperative games of strategy an agent's assumption of a choice affects his incentives. That is, incentives are relative to choices. This is the distinguishing feature of games of strategy. But furthermore, in noncooperative games a choice affects incentives through evidence about other agents' responses to the choice, not through changes in the circumstances of choice, as may happen in cooperative games. Agents can anticipate to some degree responses to their choices. This anticipation of responses is not direct. It is not, for example, a product of communication between agents as in cooperative games. It is indirect and involves reasoning. The reasoning may be strategic and go from an agent's choice to the response of other agents, or it may be nonstrategic and rest, for example, on psychological insight into the choices of other agents.

We focus more narrowly still on the noncooperative games of strategy known as *normal-form* games. Normal-form noncooperative games are generally distinguished from *extensive-form* noncooperative games. Again, the distinction fits representations of games better than games themselves. A normal-form representation of a game lists agents, strategies, and payoffs. An extensive-form representation of a game also indicates the order of moves. The representation is nondegenerate only if the game is multistage and so has a nontrivial order of moves. Strategies of a multistage game's normal-form representation are more complex than strategies of its extensive-form representation. In its normal-form representation strategies are contingency plans specifying the move to be made given various moves by other agents. Because the contingency plans, or strategies in a narrow sense of the word, are not represented by the moves composing them, normal-form representations are sometimes called "strategic-form" representations.

The normal-form representation of a game is generally proposed for single-stage games in which strategies are causally independent, or for multistage games where solutions are independent of the order of moves. The extensive-form representation of a game is generally proposed for multistage

---

[1] For a treatment of coalitions as agents in ideal cooperative games and for an exposition of the view that solutions to such games reveal collective preferences and collective utility, see Weirich (1990, 1991a, 1991b).

games where solutions depend on the order of moves. Since we treat single-stage games, we put aside extensive-form representations and the extensive-form games for which such representations are appropriate.

A normal-form representation of a game is appropriate only if the strategies of agents are causally independent. In calling a game normal-form, I imply that the strategies of agents are causally independent. This condition is usually, but not always, met in single-stage games. In a single-stage game, agents do not observe other agents' moves before making their own moves. So no agent's strategy has a causal influence on another agent's strategy via observation. In a normal-form, single-stage game it is also the case that no agent's strategy has a causal influence on another agent's strategy via any other process. Our examples concern normal-form, single-stage games of strategy. Hence they concern games where strategies are causally independent.

The normal-form games we treat are generally finite. A finite normal-form game is commonly represented by a payoff matrix, an array of ordered n-tuples, where n is the number of agents. Each number in the n-tuple indicates the payoff for one of the agents. Payoffs may be commodities or utilities. We take them as utilities, broadly conceived to include expected utilities computed with respect to the information an agent has at the end of the game. The payoff matrix has a utility n-tuple for every possible assignment of strategies to agents, that is, for every profile of strategies. An n-tuple for a strategy profile indicates the utility outcome of the profile. In some games agents are ignorant of the payoffs or outcomes of strategy profiles, and so ignorant of the payoff matrix. A payoff matrix is an objective representation of a game's possible outcomes.

The *mixed extension* of a finite game adds all probability mixtures of an agent's strategies to his list of strategies. Since the mixed extension of a finite game has an uncountable number of strategies for every agent with at least two strategies in the original game, the mixed extension cannot be represented by a payoff matrix. A payoff matrix may have a countably infinite number of strategies for an agent, but not an uncountable number. For simplicity, we work with games that have payoff matrix representations, and so put aside mixed extensions. This simplification is possible since the type of equilibrium we propose does not require mixed extensions.

For simplicity, we also restrict ourselves to games with a finite number of agents. This restriction is taken as understood throughout unless explicitly put aside. It facilitates Chapter 5's proof about the existence of equilibria. It also allows Section 2.2 to formulate idealizations about the rationality of agents without begging questions about the existence of solutions in ideal games. In general, the restriction puts aside some puzzling conflicts between intuitions about the existence of solutions in ideal games and intuitions about

11

the nature of ideal games, in which obstacles to the existence and realization of solutions have been removed.

For example, suppose a game with an infinite number of agents lacks an equilibrium. Then since an equilibrium is necessary for a solution, it lacks a solution. Can the game be ideal nonetheless? Intuition is not reliable about games with an infinite number of agents since such games are not realistic. A game in which agents have an infinite number of strategies – for instance, a mixed extension – is realistic, but not a game with an infinite number of agents.[2] Our intuition that every ideal game has a solution is secure only for games with a finite number of agents. When we say that every ideal game has a solution, we have in mind games with a finite number of agents. In fact, the intuitive concept of a solution may apply only to games with a finite number of agents; a game with an infinite number of agents may be beyond its scope. Our restriction to games with a finite number of agents enables us to avoid this treacherous territory. Our definition of equilibrium is proposed for all games. Just as we believe our new type of equilibrium suits cooperative games, we believe it suits games with an infinite number of agents. But we do not work out the details of its extension to such games because that requires dealing with complex issues independent of the nature of equilibrium.

To summarize, we consider normal-form, noncooperative, single-stage games of strategy with a finite number of agents. The categories used to pick out these games overlap. Indeed, sometimes these games are simply called normal-form games. This usage does not conform with our interpretation of normal-form games. But we take advantage of it for a short, loose description of the games we treat. We generally call them normal-form games *tout court* to abbreviate the longer, more precise description.

In normal-form games the strategies of agents are causally independent. An agent's choice does not change the circumstances of choice for other agents, and so does not change their incentives. But an agent is assumed to know the strategy he adopts, so supposition of a strategy entails supposition of his knowledge of his strategy. The supposition of an agent's strategy in deliberation therefore occasionally *carries* information that *indicates* new incentives for other agents. I say the supposition carries information rather

---

[2] A strategy is a possible decision, an intention's object. It is also an object of belief and desire and is identified by a proposition. Differences in propositions expressing strategies indicate different strategies. The strategy to play Heads in Matching Pennies differs from the strategy to play Heads at noon, even if the game is played at noon. In games in which an agent is said to have two strategies, he has two salient strategies and many other strategies equivalent to one of those two. The ideal agents we treat always have an infinite number of strategies since there are an infinite number of decisions they may make, even if a game's representation includes only the few that are salient. Realistic agents have only a finite number of strategies, however, the finite number of decisions they may make given their limited cognitive abilities.

than provides it, to emphasize that making the supposition does not alter actual information, only supposed information. And I say that supposition of the strategy indicates new incentives for other agents rather than alters their incentives to emphasize that the strategy does not cause a change in their incentives but only a change in the evidence concerning their incentives. An agent's strategy change indicates new incentives for other agents if their incentives given the agent's original strategy differ from their incentives given his new strategy. For expository convenience, I sometimes speak as though a change in an agent's strategy affects the incentives of other agents, but strictly speaking it only indicates new incentives for other agents.

Although a change in an agent's strategy merely indicates new incentives for other agents, it may alter his own incentives by providing new information about other agents' strategies. Sometimes a new strategy alters an agent's incentives, even though other agents' strategies are causally independent of his strategy, because his strategy supplies new evidence concerning their strategies, and because his incentives depend on his beliefs about their strategies.

In normal-form games agents are represented as making a single choice of a strategy, not a series of choices. The game, if finite, is represented by a payoff matrix giving the agents, their strategies, and the utility outcome for each agent of the various possible combinations of their strategies. Because these games are single-stage, agents choose independently without having observed other agents' choices, and because these games are normal-form, each agent's choice is causally independent of other agents' choices. But because these games are strategic, strategies are not evidentially independent. For at least one agent and strategy, the agent's assumption of the strategy in deliberation carries information about other agents' strategies, and so information about the outcome of the strategy. The information carried by an agent's assumption of one of his strategies is a basis for a strategic evaluation of the strategy. The assumption that a strategy is adopted may, for instance, carry information that generates incentives for alternative strategies. Under the assumption that a strategy is adopted, some other strategy may be preferred. In normal-form games strategic reasoning utilizes an agent's anticipation of the incentives his strategies generate for him.

The game of Matching Pennies is an example of the type of game we treat, an example we use repeatedly. Two agents, Row and Column, play the game. Each controls a coin. Row tries to make the coins match. Column tries to make them mismatch. The player who succeeds wins a penny, and the one who fails loses a penny. See the payoff matrix in Figure 1.4, where for each strategy profile, Row's payoff is given first. The agents make their moves simultaneously. Neither sees the other's coin until both have made

	Heads	Tails
Heads	1, -1	-1, 1
Tails	-1, 1	1, -1

Figure 1.4   Matching Pennies.

their moves. So the game is single-stage and normal-form. The agents cannot collaborate, so the game is noncooperative. Matching Pennies is also a game of strategy, provided that an agent anticipates to some degree responses to his strategies. Row, for instance, may know that if he plays Heads, Column will outwit him and play Tails.

1.3. GAMES AND THEIR REPRESENTATIONS

Having specified the type of game treated, let us now clarify the nature of games themselves. We have said that a game is a decision problem for multiple agents in which the outcome of each agent's strategy depends on the other agents' strategies. Let us also consider the ontological status of games and the features of games that determine their solutions.

We take a game to be an actual or possible concrete decision problem, not an abstract representation of a concrete decision problem. A game has a spatiotemporal location. It is represented by listing the agents, their strategies, and the payoff for each agent of each strategy profile. A concrete realization of a game's abstract representation includes details beyond those given in its representation. Since a game is specified by a representation of it – a payoff matrix in the case of a normal-form game – some take the representation itself, rather than the concrete decision problem it represents, to be the object whose solution is sought. Identifying the representation with its concrete realizations is convenient and harmless if all its realizations have the same solutions. But if different realizations of the representation have different solutions, distinguishing between the representation and its realizations is crucial, since we seek solutions to concrete games, not to their abstract representations.

Take a game of Matching Pennies. It can be represented with a payoff matrix. But realizations of the payoff matrix may differ in significant ways. Realizations of the matrix may differ, for instance, according the profile realized. In one case (H, H) may be realized. In another case (H, T) may be realized. The payoff matrix stands not for a single game but for a game type. It is a representation of each of a set of games. We do not seek a solution to the payoff matrix, but to particular concrete realizations of it.

14

These realizations may have different solutions because of differences in their unrepresented features.[3]

The standard representations of games are designed for application of the Nash equilibrium standard for solutions. Since we seek a new equilibrium standard for solutions, we allow for pertinent information not included in standard representations of concrete games. In particular, we allow for the possibility that payoff matrix representations of concrete normal-form games omit details relevant to the solutions of the concrete games. We do not assume that payoff matrices are fully adequate representations of concrete normal-form games. They may omit details of concrete games bearing on their solutions.

Matching Pennies is a game type. If *the information of the agents* in Matching Pennies changes, we obtain a different realization of its payoff matrix, and a different concrete game. But if *our information* about a realization of the matrix changes, the concrete game is unaltered, and only our resources for finding solutions to the concrete game change. We may try to find a solution to a concrete game using just the information in a standard representation of the game. If the information is insufficient to identify a solution, it may nonetheless be sufficient to identify *potential* solutions of the concrete game. Profiles that are potential solutions of a concrete game given the information in its standard representation are actual solutions, or not, depending on unrepresented facts about the concrete game. Some might regard a potential solution of the concrete game with respect to the information in its standard representation as a solution to the standard representation of the game. However, we do not examine this view. Our subject is solutions to concrete games, and not solutions to abstract representations of them.[4]

Unless we explain some other points, the close connection between games and their representations may suggest that, contrary to our view, solutions attach to abstract representations. Standard representations of games specify agents, strategies, and payoffs. A single concrete game can therefore be represented in many ways. The agents and their strategies change as we move from one representation of the game to another. For instance, a single concrete game might be represented as a cooperative game where coalitions are agents, or as a noncooperative game where only individuals are agents. In particular, a concrete game represented as a single-stage cooperative game might also be represented as a multistage noncooperative game, where offers and counteroffers are made in various stages before a final coalition structure is achieved. Furthermore, different representations of the same concrete

---

[3] Stalnaker (1994: 52–3) makes a distinction between a game and a model for a game that is similar to our distinction between a game type and a concrete game.

[4] Chapter 6, however, considers equilibria in abstract incentive structures as a means of finding equilibria in concrete games.

game may specify the same agents but partition their strategies more or less finely. For example, in Matching Pennies the partition for an agent may have one strategy, {play any side of the coin}, instead of two strategies, {play heads, play tails}. A variation in the specificity of strategies may also have consequences for the representation of payoffs. An agent's payoff given a strategy profile may be nonprobabilistic under a fine-grained partition of strategies but probabilistic under a coarse-grained partition.

Since a solution is a profile of strategies, that is, an assignment of strategies to agents, it presumes a specification of agents and strategies. We are interested in solutions to concrete games, but specify them with respect to *appropriate* abstract representations of the concrete games, in particular, appropriate listings of agents and their strategies. I do not define an appropriate representation, but a concrete game's representation is appropriate only if the implications of the representation fit the game. A game is appropriately represented as a cooperative game, for example, only if coalitions function as agents. Also, a representation of a concrete game is appropriate only if the set of strategies specified for each agent is an appropriate set for the agent to consider. Not any exclusive and exhaustive set of strategies for an agent fills the bill. For example, in the game Scissors-Paper-Stone it would be inappropriate for a typical agent to consider just two strategies: Paper, and Scissors or Stone. The disjunctive strategy is not sufficiently fine-grained. Although I do not define *appropriateness for consideration*, a set of strategies appropriate to consider includes the top contenders and other salient strategies and depends upon the cognitive powers of the agent.

When we present a concrete game by means of a representation of it, we assume that the representation, including the set of strategies for each agent, is appropriate. We trust the lore of game theory to provide appropriate representations. In our examples the strategies for an agent are the ones standardly specified, and the representations as a whole are the ones standardly specified. However, since an appropriate representation of a game may not be fully adequate for identifying solutions, we sometimes supplement a game's standard representation with additional information about the game.

Since solutions of concrete games are relative to appropriate representations of the games, a single concrete game may have solutions that vary according to its representation. The solutions may vary according to the specification of agents and strategies. For example, they may vary according to the game's representation as being cooperative or noncooperative, or as having many or few strategies per agent. Consistency, however, requires each solution of a concrete game under one appropriate representation to be compatible with some solution under any other appropriate representation. We assume that consistency holds. Solutions according to one appropriate representation of a concrete game are in fact compatible with solutions according to any other appropriate representation of the concrete game. More specifically, a change of appropriate representation, say, from cooperative to noncooperative, does not alter the profile of utilities for agents that a solution achieves. The profiles that are solutions under one appropriate representation yield the same utility outcomes as the profiles that are solutions under another appropriate representation. If a solution relative to a cooperative representation achieves a certain utility outcome of the concrete game, then some solution relative to a noncooperative representation achieves the same utility outcome of the concrete game.

Also, given the reducibility of cooperative to noncooperative games, we assume that the standards for solutions to cooperative and to noncooperative games agree with each other. In fact, given a way of carrying out the reduction, we assume that the cooperative standards for solutions are derivable from the noncooperative standards for solutions. However, we are in no position to demonstrate the consistency of solutions under alternative appropriate representations of games, or the interderivability of standards for solutions with respect to different appropriate representations of games. We first need a reduction of cooperative to noncooperative games, a full characterization of solutions, and a specification of appropriate representations. All this is beyond our project's scope. In particular, Section 1.4's definition of solutions is not precise enough to ground a demonstration of the stated assumptions about solutions.[5]

Given our assumption of consistency of solutions with respect to appropriate representations, the solutions relative to an appropriate representation may count as solutions *tout court*. Even if a solution given one representation is not a solution given another because the second representation does not have the vocabulary to express it, it is still a solution to the concrete game represented. When we say that a solution is relative to a representation, we mean that the solution presumes a representation that generates the profiles among which the solution is found. We do not mean that a strategy profile counts as solution only relative to a representation and not *tout court*. A representation supplies the terms – agents, strategies, and payoffs – with which a solution is described. A solution is relative to a representation in the way that a fact is relative to the language in which it is stated. Even though a

---

[5] The consistency of solutions relative to partitions of strategies of varying degrees of coarseness should follow from the consistency of rational options with respect to partitions of states of the world of varying degrees of coarseness. Also, the consistency of solutions relative to various exhaustive, but not necessarily exclusive, sets of strategies should follow from the consistency of rational options with respect to various sets of exhaustive, but not necessarily exclusive, options. Note that although an agent need not represent his own strategies as forming a partition, he must represent the strategies of other agents as forming a partition in order to make the states of the world that figure in his expected utility calculations form a partition.

solution of a concrete game is specified relative to an appropriate abstract representation of the game listing agents, strategies, and payoffs, it is not the same as a solution of the game's abstract representation. The solution of a concrete game under an abstract representation of the game, in contrast with a solution of the game's abstract representation, may depend on features of the concrete game omitted from its abstract representation.

Although a solution applies to a concrete game, it is itself abstract. It is a strategy profile, an intensional object. Each strategy in the profile is a possible object of decision. It is not the concrete action that issues from a decision to adopt the strategy. The collection of concrete actions the strategy profile yields has many descriptions, but not all express solutions. Some descriptions – for example, the solution is for the agents to do something – are insufficiently detailed. In general, descriptions that express solutions are descriptions of strategy profiles in appropriate representations of the concrete game. Appropriate representations have the right sort of detail for generating descriptions that express solutions.

We take an equilibrium to be a profile of jointly self-supporting strategies, in which an agent's strategy is self-supporting if he lacks a sufficient reason to switch strategy. Given this characterization, equilibria are relative to a specification of agents and their strategies, just as solutions are. Furthermore, we take being an equilibrium as a necessary condition for being a solution, and so take equilibria and solutions to appear in games of the same ontological type and to be relative to the same things. Equilibria as we understand them are thus equilibria of concrete games, just as solutions are. An equilibrium is an intensional object but concerns a concrete game. Although an equilibrium is relative to a concrete game's representation, it merely presumes the specification of agents, strategies, and payoffs given by the representation. Consistency ensures that an equilibrium relative to any appropriate representation is also an equilibrium *tout court.*

Sometimes, when there is no danger of confusion, I call a concrete game's representation a game. For instance, I call the payoff matrix representing concrete realizations of Matching Pennies a game, although strictly speaking only concrete realizations of the matrix are games. The payoff matrix itself stands for a game type. Also, speaking loosely, I call concrete realizations of the payoff matrix "versions of the game of Matching Pennies." I say, for instance, that the game of Matching Pennies has a version involving rational agents. I mean, strictly speaking, that the payoff matrix for Matching Pennies has a realization in which the agents are rational. In other words, the game type has an instance in which the agents are rational.

Another distinction concerning games also helps dispel confusion. We take games as concrete decision problems, but they can be taken in either an objective or subjective way. An *objective* game is constituted by the objective features of the agents' decision situation such as those represented by its payoff matrix. A *subjective* game is constituted by the objective game and, in addition, the subjective features of the agents' decision situation such as their beliefs about the payoff matrix. Objective and subjective games have objective and subjective solutions, respectively. Objective solutions achieve a type of success for agents; subjective solutions merely point them toward that type of success without ensuring its achievement. An objective solution depends on factors such as payoffs, whereas a subjective solution depends on factors such as preferences and rationality. Objective games with identical relevant objective features form a single objective game type and have common objective solutions. Likewise subjective games with identical relevant subjective features form a single subjective game type and have common subjective solutions. Subjective games of the same objective game type are distinct if the agents' information about the objective game type differs from one subjective game to another. For instance, agents play different subjective versions of the objective game type Matching Pennies if in one version they know the payoff matrix but in another version they do not.

There are different sorts of objectivity. We are concerned about objectivity of information, not value. Objective utilities on our view are just fully informed subjective utilities. They diverge from subjective utilities when agents are not fully informed. Similarly, objective and subjective solutions diverge when agents are not fully informed. On our view an objective solution has a subjective component. An objective solution is defined in terms of payoffs, and these are utilities for amounts of commodities rather than amounts of commodities.

We treat subjective games since we are interested in bringing principles of rational choice to bear on decisions in games, and since rational choice depends on subjective features of agents such as information. We say that two groups of agents are in different game types if the groups' information about the payoff matrix differs, even if the groups' situations are in fact governed by the same payoff matrix. The subjective solutions of their subjective games may differ even if the objective solutions of their objective games are the same. Objective and subjective solutions and game types come together, however, in ideal games where agents are fully informed about the objective features of their games, and rational preferences are shaped by payoffs and the like. In those games there is just one subjective game type for each objective game type, and the subjective and objective solutions coincide. This is fortunate. The distinction between objective and subjective games is vague since the distinction between objective and subjective features of agents' decision situations is vague. But the vagueness does not cause any problems for our main project. That project concerns ideal games, and the objective and subjective games in a concrete decision situation have consistent objective and subjective solutions given our idealizations.

The distinction between objective and subjective games helps clarify points about the completeness of descriptions of games. What may at first seem to be an incomplete description of a game type may be a complete description of a subjective game type, that is, a specification of everything needed to determine subjective solutions, or profiles of jointly rational strategies. The description may incompletely specify the payoff matrix, for instance. But it may nonetheless completely determine the profiles of strategies that are jointly rational. Rationality for the agents takes account of their information, so their choices may be jointly rational even if they are ignorant of the payoff matrix. A description insufficient for determining the payoff matrix or objective solutions may be sufficient for determining the jointly rational strategies and subjective solutions.

1.4. SOLUTIONS DEFINED

The concept of a solution to a game is a theoretical concept. So our definition of a solution is guided by theoretical considerations as well as established usage. Our definition conforms to usage of the term *solution* in game theory. But established usage is not a sufficient guide in matters of detail. As a guide to details, we use theoretical considerations that may be controversial.

We are interested mainly in subjective solutions, profiles of jointly rational strategies. But objective solutions are the foundation of our study of subjective solutions. Basic intuitions about solutions are about objective solutions. Intuitions about subjective solutions are not basic, but derivative from intuitions about rationality. Our basic intuitions about objective solutions, being independent of our basic intuitions about rationality, are an important resource. We can use basic intuitions about objective solutions, and the agreement of objective and subjective solutions in ideal games, to learn about subjective solutions and to refine principles of rationality. Intuitions about subjective solutions, in contrast, are dependent on intuitions about principles of rationality and so do not provide a means of refining those principles. We do not have basic intuitions about subjective solutions independent of basic intuitions about rationality, as we do for objective solutions. So we cannot make significant discoveries about rationality using intuitions about subjective solutions as our foundation.

The theoretical importance of objective solutions helps explain why our interest in subjective solutions, or *joint* rationality, is confined mainly to ideal games. In ideal games joint rationality is tied to basic intuitions concerning objective solutions. Mere rationality does not have the same connection with objective solutions. Ideal games give us independent insight into joint

20

rationality but not into mere rationality. On the other hand, in nonideal games joint rationality loses its connection with objective solutions, and mere rationality becomes more interesting because it is more fundamental. Mere rationality has more theoretical interest than joint rationality when joint rationality is untethered to objective solutions and so not a beneficiary of independent direction.

We seek a definition of subjective solutions but use intuitions about objective solutions in ideal games to guide the definition. Our investigation of subjective solutions starts with the view that a solution is a profile of jointly rational strategies. Since there are many ways of interpreting joint rationality, there are many plausible definitions of a solution. We use the theoretical role of a solution to decide among them. A solution is to be the outcome of strategic reasoning in games of strategy of all types, both ideal and nonideal. We want to define joint rationality so that this turns out to be the case and, in addition, so that in ideal games a solution achieves objective success.

One view takes joint rationality to be universal rationality. According to it, a solution is a profile of rational strategies, that is, an assignment of strategies to agents such that each strategy is rational for the agent to whom it is assigned. This view takes the strategies in a solution to be rational with respect to actual circumstances – actual opportunities, beliefs, and desires. As a result, it fails to produce what we intuitively think of as a solution to a game. Take the normal-form game in pure strategies in Figure 1.5. Column has a dominant strategy, C1. Given C1, Row's best strategy is R2. We think of (R2, C1) as the solution to the game. But suppose that Column irrationally chooses C2, and Row foresees this. Then Row's rational strategy is R1. Column's rational strategy remains C1. So if we form a profile of strategies that are rational with respect to actual circumstances, including Column's foreseen irrationality, we obtain the profile (R1, C1). This is not a solution in the intuitive sense. It is not a solution according to the intuitions about strategic rationality that we are trying to capture. Even if the strategies in the profile are rational, they are not jointly rational.

Another problem with defining a solution as a profile of strategies rational with respect to actual circumstances appears in games that intuitively have multiple solutions. Take the ideal normal-form game in Figure 1.6. (R1, C1)

$$0, 2 \qquad 2, 1$$

$$2, 2 \qquad 1, 1$$

Figure 1.5   Universal rationality without joint rationality.

21

$$1, 1 \qquad 0, 0$$

$$0, 0 \qquad 1, 1$$

Figure 1.6   Multiple solutions and universal rationality.

and (R2, C2) are both solutions in an intuitive sense. But if the agents actually realize (R1, C1), then, being ideal and so prescient, each agent has only one rational choice. It is R1 for Row and C1 for Column. Hence there is only one profile of strategies rational with respect to actual circumstances, namely, (R1, C1). The strategies of an agent that are rational given actual circumstances are limited by what other agents actually do. Rationality with respect to actual circumstances does not provide for multilateral changes of strategy that achieve an alternative solution. Taking a solution as a profile of strategies rational with respect to actual circumstances does not allow for the possibility that a game has several solutions.

Let us consider another definition of a solution, which states that a solution issues from the simultaneous rationality of agents. More precisely, it says that a solution is a profile realized if all the agents decide rationally. Simultaneous rationality is not the same as universal rationality. Universal rationality is with respect to actual circumstances. The simultaneous rationality of agents is counterfactual if some agents in fact decide irrationally. The strategies rational given the counterfactual supposition of simultaneous rationality may not be the ones rational given actual circumstances. In particular, if agents are prescient but some are irrational, a counterfactual supposition of simultaneous rationality may change the strategies that are rational.

Defining a solution in terms of simultaneous rationality instead of universal rationality handles cases in which some agents are irrational. But cases with multiple solutions defeat the revision. Although in some concrete games several profiles are intuitively solutions, at most one profile is realized if all the agents decide rationally. For instance, take the ideal normal-form game in Figure 1.7.[6] Suppose that in a concrete realization of the payoff matrix the agents break ties in ways that favor their first strategies. Then the profile realized if all decide rationally is (R1, C1). This profile is the only profile that is realized if all decide rationally. No other profile has this distinction. Hence (R1, C1) is the only solution according to the revised definition. The other profiles are ruled out although intuition tells us that every profile is a solution, even for concrete realizations of the payoff matrix, as opposed to the payoff matrix itself. In a concrete realization of the payoff

[6] It is appropriate for a game's representation to distinguish tying strategies rather than conflate them since tie breaking is an important part of strategic reasoning.

$$1, 1 \qquad 1, 1$$

$$1, 1 \qquad 1, 1$$

Figure 1.7   Multiple solutions and simultaneous rationality.

matrix in which ties are broken in favor of R1 and C1, R2 and C2 are still rational choices. Every profile has jointly rational strategies. Every profile is intuitively a solution.

To obtain a satisfactory definition of a solution, let us view the joint rationality of strategies in a solution as a type of conditional rationality. Let us say that the strategies in a profile are jointly rational if and only if each is rational *given* the profile. This leads to the following definition of a solution:

A *solution* is a (feasible) profile of strategies that are rational given the profile's realization, in other words, a (feasible) profile of strategies that are rational if realized together.

In normal-form games all profiles are *feasible*, or possible outcomes of the game, so the explicit restriction to feasible profiles may be dropped. Notice that the condition specified by the definition is realization of a profile, not knowledge of its realization. The two conditions are equivalent only if agents are prescient so that for any profile, given that it is realized, agents know that it is realized.

According to our definition, a solution is a profile of strategies that are rational given that the profile is realized. When it is important to distinguish a solution of this type from solutions of other types, I call it a *relative* solution. I call it relative since the rationality of the strategies in a solution of this type is relative to a profile, or conditional with respect to the profile. The strategies a solution comprises are appraised with respect to the incentives that hold if the strategies are adopted.

Relative solutions differ from profiles of rational strategies, or *nonrelative* solutions. The strategies constituting nonrelative solutions are rational with respect to actual circumstances, not with respect to counterfactual profiles. Differences between relative and nonrelative solutions arise in various ways. A profile of jointly rational strategies, if realized, has strategies that are rational, period. But not every profile of jointly rational strategies is realized. Some unrealized profiles of jointly rational strategies comprise strategies that are not rational, period. For instance, an agent's strategy in an unrealized profile may be rational given the profile but not rational, period, because the agent believes that the other agents do not adopt their parts in the profile.

Relative and nonrelative solutions have an interesting relationship in ideal games. In ideal games divergences due to irrational agents disappear. The

23

profile realized is a relative solution. With respect to that profile, the strategies in it are rational. For that profile, the circumstances with respect to which the relative rationality of its strategies is assessed are the actual circumstances, with respect to which the nonrelative rationality of its strategies is assessed. Hence the profile realized is both a relative and a nonrelative solution. However, even in ideal games divergences due to multiple solutions remain.

On the other hand, in certain types of *nonideal* games relative and nonrelative solutions coincide. Unless a game involves some degree of prescience so that the condition that a profile is realized carries some relevant information, rationality given a profile is just rationality. Then there is no difference between relative and nonrelative solutions.

Relative solutions are the topic of classical game theory. We concentrate on relative solutions because they agree with objective solutions in ideal games. Moreover, relative solutions register intuitions about strategic rationality by letting the appraisal of the strategies in a solution take account of the information carried by supposition of the strategies. There is also a practical reason for addressing relative solutions. In an ideal game finding relative solutions requires less information about the game than finding nonrelative solutions. Given prescience, the condition with respect to which a solution is relative carries information that may compensate for a shortage of information about the game.

Although nonrelative solutions are not the solutions game theorists generally have in mind, these solutions are independently interesting. They are interesting to the decision theorist, who wants primarily to know which strategies for an agent in a game are rational taking circumstances as they are, not given the realization of a perhaps counterfactual profile. Although relative solutions are our principal interest, we occasionally consider nonrelative solutions.

1.5. CONDITIONAL RATIONALITY

Our definition of a solution needs further explication. It interprets joint rationality as conditional rationality. We have to say what we mean by a strategy's being *rational given a condition*.

To begin, let us distinguish conditional rationality from conditional assessment of a strategy. Although conditional rationality involves a type of conditional assessment, it is distinct from many types of conditional assessment of a strategy. Conditional rationality is rationality with respect to a condition. Conditional assessment of a strategy is any sort of assessment with respect to a condition. Failure to keep this in mind may lead to confusion. Self-support, a requirement of nonconditional rationality, involves

24

a conditional assessment of a strategy. The nonconditional rationality of a strategy depends on the utility ranking of the strategy given that it is adopted. But the rationality of a strategy with respect to one set of circumstances does not depend on its rationality with respect to another set of circumstances. It is not the case that a strategy is rational partly because it is rational on the assumption that it is adopted. Such dependency would generate an infinite regress. Rather, the rationality of a strategy depends on a certain conditional assessment of the strategy. A strategy is rational partly because it meets certain standards on the assumption that it is adopted – in particular, sufficient reasons to switch are absent.

Another important distinction concerns the method of evaluating a strategy for rationality given a profile. There are two ways of conducting the conditional evaluation: (1) It can be done simply under supposition of the condition, or (2) it can be done under supposition of the condition after excusing or putting aside any irrationality the condition involves. According to the second method of evaluation, an option is a rational choice given an agent's circumstances if and only if the option is rationally chosen by the agent, accepting or taking for granted his circumstances whatever they may be, even if they are fashioned by his mistakes. To illustrate the difference between the two methods of conditional evaluation, take the condition that an agent fails to consider his top option. According to the second method of conditional evaluation, an option is a rational choice given the condition if it is the most choiceworthy of the options the agent does consider. His failure to consider his top option is excused during the evaluation. But according to the first method of conditional evaluation, only his top option is a rational choice given the condition; failure to consider that option does not lower the standards of rationality for his choice.

The two types of conditional rationality differ according to whether the evaluation accepts as a point of departure the agent's circumstances under the condition, whatever they may be, even if they include some mistakes by the agent. The second type accepts the agent's circumstances and then applies standards for a rational way of proceeding in those circumstances. It is forward-looking and focuses on the causal consequences of options in the agent's circumstances. It considers ways of compensating for the agent's mistakes. The first type of conditional rationality evaluates options without concessions to prior mistakes. It allows for the possibility that a decision is deficient, not because of the choice made to reach it, but because of the beliefs, desires, deliberations, and so on, that set the stage for the choice. It does not focus exclusively on the causal consequences of options; it is not purely forward-looking. It looks at the history of the agent's current decision situation, not just the final choice. Its topic is overall achievement, not compensation for error. It does not use compensatory standards of evaluation. It uses the noncompensatory standards we apply when we say that an agent's choice of a career is irrational because he did not consider a wide enough variety of careers, or when we say that an agent's decision to travel by train is irrational because it springs from an irrational fear of flying.

The two types of conditional rationality are related. Given a condition, an option is rational in the first, noncompensatory way only if it is rational in the second, compensatory way. An option is conditionally rational not putting aside mistakes involved in the condition only if it is conditionally rational putting aside mistakes involved in the condition. Also, for an ideal agent, who enters a decision problem without having made any mistakes, an option rational in the compensatory way is also rational in the noncompensatory way.

When defining solutions in terms of conditional rationality, which type of conditional rationality is appropriate? In saying that a profile is a solution if and only if each strategy is rational given the profile, which type of conditional evaluation should be adopted? Do compensatory or noncompensatory standards work best? Should we excuse mistakes involved in the condition or not? Since the profile that serves as a condition contains the strategies to be evaluated with respect to it, we do not want to excuse the profile, and along with it any mistaken strategies it contains. We do not want a strategy in a profile to be excused under supposition of the profile simply because it is in the profile. This would lead to a mistaken strategy, in effect, excusing itself. We therefore adopt the first, noncompensatory type of conditional rationality. Taking a profile as a condition supplies a supposition under which noncompensatory standards of evaluation are applied.

This interpretation of conditional rationality thwarts straightforward argument from the existence of a rational choice in every decision situation to the existence of a solution in every game. The absence of dilemmas of rational choice implies only the existence of an option that meets compensatory standards of rationality. If an agent enters a decision problem with irrational beliefs and desires, every possible decision may fall short of pertinent noncompensatory standards of rationality. Therefore, to support the existence of solutions, we rely on idealizations that bring together compensatory and noncompensatory standards of rationality.

To clarify the definition of a solution, we have explained the way the supposition of a profile is to be used in evaluating a strategy. We must also explain the method of supposition that is to be applied to a profile. Our definition of a solution asks whether all the strategies in a profile are rational given the realization of the profile. If the profile is not actually realized, the supposition that it is realized is contrary-to-fact. Since there are two main kinds of contrary-to-fact supposition – indicative and subjunctive – we have to say which kind of supposition is intended.

Both indicative and subjunctive supposition can be explicated by the relations of nearness among possible worlds used to explicate counterfactual conditionals. Supposition of a proposition involves supposition of the nearest world in which the supposition is true (we assume there is such a world). Types of supposition differ according to the nearness relation involved. Indicative supposition is guided primarily by epistemic considerations. Under indicative supposition of a profile, the information with respect to which choices are assessed changes to accommodate the information carried by supposition of the profile. In contrast, subjunctive supposition is guided primarily by causal considerations. Under subjunctive supposition of a profile, the causal circumstances with respect to which choices are assessed change to accommodate the causal consequences of the profile.

We adopt indicative supposition for the definition of a solution. The main reason is related to the definition's purpose. We want to define a solution in a way that makes a solution the outcome of good strategic reasoning, in particular, in normal-form games in which strategies are causally independent. Good strategic reasoning in normal-form games responds to evidence choices provide. It keeps causal circumstances constant but allows variations in information supposed given variations in the profile supposed. Now a profile of strategies that *would be* rational if the profile *were* realized is evaluated with respect to counterfactual causal circumstances, not the actual causal circumstances, when the profile is counterfactual. But strategic considerations in normal-form games require an evaluation of all profiles with respect to actual causal circumstances. They require an evaluation of each profile with respect to the hypothetical information the profile carries, but also with respect to the actual causal circumstances. Strategic reasoning in normal-form games entertains changes in the epistemic rather than the causal situation as the profile supposed changes. The indicative supposition of a profile registers the profile's evidential impact rather than its causal impact. It provides causal constancy and allows evidential variation, as strategic evaluation requires. Indicative supposition takes account of strategic considerations, so strategic reasoning uses indicative supposition of profiles. To make solutions yield the outcome of good strategic reasoning, therefore, we define a solution as a profile of strategies that *are* rational if, perhaps contrary-to-fact, the profile *is* realized. In other words, a solution is a profile such that if it *is* realized all the strategies in it *are* rational. Because of the distinctive role of evidential considerations in strategic reasoning in normal-form games in particular, and games of strategy in general, indicative supposition is best for our definition of a solution.

Picking indicative supposition of a profile for the definition of a solution has an additional, theoretical advantage. Chapter 4's account of strategic reasoning uses a decision rule that evaluates an option given the information

27

carried by the assumption that the option *is* realized. Since this decision rule uses indicative supposition, it is easier to derive conclusions about solutions if the definition of a solution uses indicative supposition as well. In this case we obtain conformity between types of supposition. We can use the decision rule to argue for the joint rationality of the strategies in profiles that are solutions according to the definition. Without conformity of types of supposition, we could not easily argue that the strategies in solutions are the product of rational strategic reasoning. In fact, our decision rule would argue against our definition of a solution as an account of the intuitive concept of a solution.[7]

Since our definition of a solution uses indicative supposition of a profile, it is equivalent to another common way of defining a solution. Jointly rational strategies are strategies rational taken together. So in a profile of jointly rational strategies, each agent's strategy is rational *given* strategies that are rational for the other agents. The joint rationality of the strategies therefore amounts to the rationality of each strategy *given* the other strategies.

---

[7] An agent's having a rational strategy means that he has a strategy such that if he *adopts* it, he satisfies the relevant standards of rationality. Here the supposition of the strategy is indicative. The relevant standards, however, may involve subjunctive supposition of the strategy. The standards may concern what *would* happen if the strategy *were* realized. A rational strategy in a decision problem may not be realized. Then it may be that if the agent realizes the strategy, he is in a different decision problem and the strategy is not rational. Hypothetical conclusions about the rationality of the strategy are sensitive to the order of suppositions and, in particular, to whether the first supposition is the decision problem or the strategy.

2

Idealizations

In an ideal game each agent is ideal and ideally situated. The agents are rational, prescient, and informed about the game. These are strong idealizations, and much recent work in game theory is sensibly concerned with weakening them, especially the idealization of prescience. We keep the idealizations at full strength, however, because of their role in explications of important features of strategic reasoning. This chapter explains and motivates our idealizations.

2.1. THE CRITERION FOR IDEALIZATIONS

The agents in an ideal game decide rationally, have the power to anticipate each other, and have full knowledge of the payoff structure of the game and the circumstances of the agents. The foregoing is only a rough characterization of ideal games, however. It leaves unsettled some matters resolvable in ways that either support or undermine our claims about solutions and equilibria in ideal games. It is tempting to support or undermine those claims with stipulations about the features of ideal games, but this temptation should be resisted.

To support our claim that solutions exist in ideal games, it is tempting to stipulate that any game without a solution is not ideal. To undermine our claim that Nash equilibria are absent in some ideal games, it is tempting to stipulate that a game is ideal only if each agent is prescient, is informed about the game, and adopts an incentive-proof strategy. Since this condition is not met in games without Nash equilibria, such games would then be classified as nonideal. It is, however, a mistake to use stipulations about ideal games in this way to shore up or undermine arguments against the standard of Nash equilibrium. It is more fruitful for game theory to take ideal games to be defined independently of solutions and specific standards of rationality. Then features of ideal games may be used to explain features of solutions and to verify standards of rationality. So let us resist the temptation to resolve issues about ideal games by stipulation. Let us treat the concept of an ideal game as primitive. Let us flesh it out by considering its theoretical role.

Here are some general points about the role of the idealizations. The idealizations remove obstacles to the existence and realization of *objective*

29

solutions. Moreover, *subjective* solutions as we have defined them coincide with objective solutions in ideal games. These two basic points give content to the theoretical concept of an ideal game. They give us a handle on the connection between idealizations and subjective solutions. Idealizations are more closely connected to objective solutions than to subjective solutions, but they are connected to subjective solutions through the connection between objective and subjective solutions in ideal games. Idealizations make it possible for subjective solutions to meet objective standards for solutions. They remove obstacles that prevent agents' choices from meeting objective standards. They move rational choices closer to successful choices.

An idealization is not simply a restriction, but a restriction that renders idle one of the factors in the explanation of rational strategies. Explanations of rational strategies take account of extenuating circumstances that permit rational strategies to fall short of objective standards. Idealizations simplify these explanations by putting excuses aside. The idealizations of prescience and full information concerning the payoff structure, for instance, remove epistemic limitations that provide excuses for substandard choices. They allow explanations of rational choices to ignore the effect of epistemic limitations. On the other hand, our assumption that the number of agents is finite is a restriction, not an idealization. It simplifies an account of rational strategies by restricting the cases considered rather than by controlling for a factor in the explanation of the rationality of strategies.

Ideal games are ones without excuses for failing to meet objective standards. Thus a full account of the features of ideal games presumes a full specification of objective standards for choices in games, something beyond the scope of this book. To simplify, we rely on a general understanding of the nature of ideal games and a specification of their salient features in the case of normal-form games. The agents' rationality, informedness, and prescience are standard idealizations for normal-form games. They remove the principal excuses for failing to meet objective standards. However, additional idealizations may be needed to remove other excuses. Suppose, for instance, that the payoff matrix for a normal-form game has a realization where agents are rational, informed, and prescient, but there is no solution. I do not think that there is such a case, but suppose one exists. Before abandoning our assumption that every ideal game has a solution, I would consider whether some additional idealization's absence explains the absence of a solution.[1]

---

[1] I believe that every payoff matrix for a normal-form game has a realization that is an ideal concrete game. It seems possible to obtain such a realization by making the agents rational, informed, and prescient. But faced with a case like the one described, I may be forced to abandon this belief. Perhaps it will be impossible to supply additional idealizations to remove excuses for the absence of a solution. Then I may conclude that the payoff matrix lacks an ideal realization. I will draw this conclusion before conceding that some ideal game lacks a solution.

2.2. THE IDEALIZATION ABOUT RATIONALITY

To avoid begging the question about the existence of solutions in ideal games, I characterize ideal games in a way that respects the independence of the concept of an ideal game and the concept of a solution. Since a solution is a profile of jointly rational strategies, the crucial idealization concerns the rationality of the agents. I fill out the idealization in a way that leaves open the question of the existence of solutions in ideal games. I let the idealization of rationality be that agents are in fully rational states of mind, with rational beliefs and desires, and make jointly rational choices *if possible*.

According to our idealization, *agents are jointly rational if it is possible for them to be rational jointly.* In other words, the agents realize a solution if a solution exists. We hold that every ideal game has a solution; the agents can realize a profile in which strategies are jointly rational. But we do not make the existence of a solution part of the idealization. Idealizations control for factors that explain the existence of a solution and so do not include the existence of a solution. The existence of a solution in an ideal game should be demonstrated, not simply stipulated. Although a demonstration of the existence of a solution is beyond this book's scope, we do plan to demonstrate the existence of an equilibrium, a necessary condition for the existence of a solution. We do not make the existence of an equilibrium in an ideal game a matter of stipulation.

Our idealization loosens the connection between rationality and ideality for individual agents. Rationality requires pursuit of sufficient incentives to switch strategy. But the definition of an ideal agent does not stipulate that an ideal agent pursues sufficient incentives to switch strategy. To be ideal, an agent need only pursue sufficient incentives to switch strategy where joint pursuit of sufficient incentives is possible. If an agent fails to pursue a sufficient incentive where exactly one exists or, where there are several, fails to pursue any of them, he is irrational, but not necessarily nonideal. Where pursuit of sufficient incentives by all agents is impossible, all may be ideal without all being rational.

As we understand the idealization, it makes joint rationality robust. The idealization is not simply that joint rationality is achieved if possible. Rather joint rationality is achieved if possible *given any situation*.[2] The agents are jointly rational in any situation, factual or counterfactual, where they can be jointly rational. The idealization eliminates all profiles where agents are not rational given the profile, provided that some profile allows all agents to be rational given the profile. The robustness of joint rationality holds for subsets of agents as well as the whole set of agents. In virtue of the idealization, if

---

[2] For more on robust rationality by another name, see Sobel (1994, Chp. 16) on "resilient" rationality.

an agent adopts a strategy, the other agents adopt a response that is jointly rational if possible.

Although our idealization makes joint rationality robust, it does not make joint rationality inevitable. Even in ideal games where joint rationality is possible, it is not achieved given certain conditions. For instance, given a full profile with strategies irrational given the profile, joint rationality is not achieved. The idealization is not so powerful that it makes the agents' knowledge change under supposition of a profile so that their strategies are rational given the profile. Where supposition of a profile generates conflict between the idealization of joint rationality and the idealizations of prescience and informedness, the latter have priority over joint rationality. Prescience and informedness are maintained at the cost of the agents' joint rationality.[3]

Suppose that joint rationality is possible in a game. The idealization about agents ensures that they make rational choices in the profile actually realized, but it does not ensure that they make rational choices in every profile supposed. Given some profiles, some agent's choice may be irrational. Although the agents are actually rational, they are not necessarily rational. Their rationality may vanish in profiles hypothetically supposed. However, their individual rationality is robust. If the profile supposed involves an irrational strategy, it is an isolated mistake and the agent is rational aside from that mistake insofar as possible, given the priority of prescience and informedness. As Chapter 5 shows, the robustness of an agent's rationality is important in computing an agent's responses to strategies of other agents, and the incentives of other agents given those responses.

We use the indicative mood to explicate the possibility of joint rationality. When we say that joint rationality is possible, we mean that there is a profile such that if it *is* realized, then the agents *are* jointly rational. We do not mean that there is a profile such that if it *were* realized, then the agents *would be* jointly rational. We explicate the possibility of joint rationality using indicative supposition to achieve conformity with the type of supposition relevant to solutions; the idealization is supposed to set the stage for the realization of a solution. Using indicative supposition also makes our explication of the possibility of joint rationality conform with our explication of the possibility of individual rationality. When we claim that it is possible for an agent to be rational, we mean that there is a strategy such that if the agent *adopts* it, the agent *is* rational. The conformity of our explications creates agreement between rationality and joint rationality when a profile of rational strategies is realized.

---

[3] Stalnaker (1994: 66–8) suggests similar priorities for idealizations under counterfactual suppositions.

32

To understand the motivation for the idealization about joint rationality, it is important to distinguish, as in Section 1.1, between the existence and the realization of solutions. The existence of solutions is governed by a game's nature, and the realization of solutions is governed by the performance of agents in the game. The two phenomena are related. The realization of a solution obviously requires the existence of a solution. But also the existence of a solution may depend upon the performance of agents. The existence of a solution requires a profile of jointly self-supporting strategies, and self-support depends on responses to strategies, and so the performance of agents. Hence an idealization for the existence of solutions may treat agents' performance. The idealization that agents are jointly rational if possible treats performance but bears on the existence of solutions.

To support our view that ideal games have solutions, Chapter 5 derives the existence of equilibria, a necessary condition for the existence of solutions, from idealizations such as informedness, prescience, and achievement of joint rationality if possible. The idealization that agents are jointly rational if possible bears on the existence of solutions but does not beg the question of the existence of solutions. It does not assume that solutions exist. Without begging the question, we can assume the realization of profiles of jointly rational strategies *if* they exist. We only beg the question if we assume without the proviso that agents realize jointly rational strategies.

How do agents achieve joint rationality in ideal games where it is possible? Superficially, the explanation is that each agent is rational. For if each is rational, then each is rational given the profile realized, and so they are jointly rational. But the agents' knowledge of their situation and the rules for equilibrium selection also play a role in a full explanation of the achievement of joint rationality, or the realization of a solution. Since we do not formulate rules for equilibrium selection, we do not try to explain fully how a solution is realized. Fortunately, in order to support our claims about the existence of solutions, we do not have to explain how, in cases where it is possible, agents realize a solution.[4]

Is the assumption that agents are jointly rational if possible an appropriate idealization and not just a restriction? As Section 2.1 indicates, whether an assumption is justified as an idealization depends on the nature of an objective solution. Some idealizations are justified for one type of objective solution but not for another. For example, different types of prescience are appropriate for relative and nonrelative objective solutions. We propose the assumption that agents are jointly rational if possible as an idealization for relative objective solutions. These are objective solutions that have profiles

---

[4] Are there rules for selection of a solution similar to rules for selection of an equilibrium? Do concrete games have multiple solutions or just one solution favored by rules for selection of a solution? We put such questions aside.

of jointly rational strategies as their subjective counterparts. A basic intuition about these objective solutions is that they entail each agent's rationality with respect to the solution.

The assumption that agents are jointly rational if possible is an appropriate idealization for relative objective solutions. Idealizations remove obstacles to the realization of an objective solution. One way to remove all obstacles to relative objective solutions in one sweep, without presuming the existence of a solution, is to let the idealizations include the assumption that agents are jointly rational if possible. Then in an ideal game the agents realize a relative objective solution if one exists, for they realize a relative subjective solution if one exists, and given the other idealizations this is a relative objective solution. Where joint rationality is possible, a failure to attain it is an obstacle to the realization of a relative objective solution. We can be confident of this even without a full account of relative objective solutions. The idealization that agents are jointly rational if possible thus removes an obstacle to the realization of a relative objective solution, and so is justified.

The idealization about joint rationality may be more powerful than necessary for the existence of solutions. It is powerful enough to ensure the realization of a solution wherever one exists, and so powerful enough to treat the realization of solutions as well as the existence of solutions. However, it does not hurt to use the idealization in dealing with the existence of solutions, since the idealization, even if powerful, does not beg the question of the existence of solutions.

Our idealization concerns the agents in a game taken together instead of individually. Is it possible to reformulate the idealization so that it applies to each agent individually instead of to all taken together? The idealization about joint rationality is not collectivistic, since joint rationality involves conditional rationality rather than collective rationality. But it is nonetheless desirable to derive the idealization from deeper idealizations about the behavior of individual agents. Moreover, it is desirable to derive the idealization from general idealizations for individuals that describe behavior even in games where it is impossible for all to be jointly rational. Although our idealization's lack of generality does not matter in ideal games, where, as we argue, joint rationality is possible, it is theoretically attractive for an idealization to be independent of other idealizations and so be general and have consequences for nonideal games. Can our idealization about joint rationality be derived from other, more fundamental idealizations? Can the idealization that agents are jointly rational if possible be derived from idealizations such as the rationality of agents?

From the rationality of agents, we can, in fact, easily derive that they are jointly rational if possible. The rationality of agents entails their rationality given the profile realized, and hence their joint rationality. So if

34

agents are rational, they are jointly rational, and therefore jointly rational if possible. Robust rationality similarly entails robust joint rationality where possible. But the idealization of universal rationality is too strong given our aim of adopting idealizations that make the existence of a solution in an ideal game an open question. The existence of a solution follows directly from universal rationality, since universal rationality entails joint rationality and thus a solution. So we forgo the idealization that each agent is rational.

What about the weaker idealization that each agent is rational if it is possible for him to be rational? It entails that each agent is rational given our view that individual rationality is always possible, and so yields the idealization about joint rationality. Unfortunately the weaker idealization is still too strong since in entailing universal rationality it also entails the existence of a solution. Our idealization about joint rationality does not entail that a solution exists in an ideal version of Matching Pennies. If, however, we had said that each agent is rational if it is possible for him to be rational, then given our view that it is always possible for an agent to be rational it would follow that Matching Pennies has a solution. It would follow that all agents are rational, and so rational given the profile realized, and so jointly rational. And if agents realize a solution, then, of course, one exists. Given our claim that individual rationality is always possible, the idealization that each individual is rational if it is possible for him to be rational begs the question about the existence of a solution.

Other basic idealizations about individuals may avoid the problem. They might allow us to derive the idealization that joint rationality is achieved where possible and yet not be too strong. One of the main difficulties, however, is the following: A derivation of our idealization from more basic idealizations would have to show, in particular, how without communication agents achieve joint rationality in an ideal normal-form game where joint rationality can be achieved in several ways. The achievement of joint rationality seems to call for coordination in games where several profiles have jointly rational strategies. We have to show that basic idealizations lead all agents to a single profile where they are jointly rational. We have to show that it is not the case that for each agent, his rationality leads to a profile where some other agent fails to be rational. Searching for a demonstration that overcomes this problem takes us too far afield. Deriving the idealization about joint rationality from more fundamental idealizations goes beyond the scope of our project.

Moreover, we do not need basic idealizations like the idealization that each agent is rational, or rational if possible, to enable agents to predict each other's choices for the purpose of strategic reasoning. Another idealization, prescience, gives agents foreknowledge of the choices of other agents. Given that strong idealization we can make do with a weaker idealization about agents' rationality, one that does not entail the existence of a solution.

In any case, even if the idealization about joint rationality can be broken down into more basic idealizations about the rationality of agents, it is plain that the main consequence of those more basic idealizations is that agents are jointly rational if possible. That is the upshot of idealizations concerning each agent's deliberations about other agents, including their deliberations about his deliberations and the like. So without laying out more basic idealizations from which it derives, we can take our idealization to be that agents are jointly rational if possible.

For the foregoing reasons, we rest content with our idealization about joint rationality. We point out only a slightly more fundamental formulation of the idealization. Our idealization is derivable from the idealization that the agents are *all rational* if joint rationality is possible. The latter idealization concerns universal rationality with respect to a condition, the same condition as in our idealization concerning the achievement of joint rationality. The new idealization is more fundamental, however, since it individualizes the rationality of agents given the condition instead of attributing rationality to them jointly.

The new idealization is equivalent to the old. We can show this by showing the equivalence of universal rationality and joint rationality, assuming that the equivalence survives if we add the condition about the possibility of joint rationality, and likewise if we add that universal and joint rationality are robust. The equivalence of universal rationality and joint rationality concerns the actual case, not counterfactual cases. That is, it obtains with respect to the profile realized, not with respect to unrealized profiles. As Section 1.4 shows, universal rationality and joint rationality may diverge with respect to profiles not realized.

To see that universal rationality and joint rationality are equivalent, consider the following: If all agents are rational, then they are rational given the profile they realize, and so are jointly rational. Also, if they are jointly rational, then they are rational given the profile they realize, and so are all rational. Universal rationality is universal rationality given the profile realized, and so is joint rationality. Agents are universally rational if and only if they are jointly rational.

The equivalence does not require prescience. Suppose that agents are not prescient. Then suppose that an agent's strategy in the profile realized is rational although he is mistaken about the profile realized. His strategy is still rational given the profile, since supposition of the profile does not give the agent knowledge of the profile. His mistake about the profile realized endures given supposition of the profile.

2.3. PRESCIENCE

Prescient agents have the power to anticipate others' choices. We take ideal agents to have this strong predictive power for three theoretical reasons: First, strategic reasoning in its purest form arises when an agent's choice provides certainty about the choices of other agents. The idealization of prescience brings problems of strategic reasoning to the fore, without attendant problems of uncertainty. Second, the idealization makes objective and subjective solutions coincide so that intuitions about objection solutions can be used to refine intuitions about subjective solutions and underlying principles of rational choice. Third, to be general, a normative theory of games must encompass cases where agents are prescient even if such cases are fictitious.

Prescience has many precedents in the literature. For example, von Neumann and Morgenstern (1944: 105, 146–8) assume that in ideal games agents "find out" the strategies of other agents, even in hypothetical situations. Sometimes, in articles where prescience seems absent, idealizations about common knowledge implicitly generate a form of prescience. Where prescience is altogether absent, the topic is usually different from mine. It may, for example, be games where inductive rather than strategic reasoning is crucial for rational choice.

As an idealization prescience removes certain obstacles to equilibria and solutions, but it also makes the existence of equilibria more difficult to establish in some cases (see Section 5.2.1). Some theorists may think that prescience creates intractable problems of strategic reasoning. One of my goals is to show that the problems can be resolved, and that prescience is a useful idealization.

As are other idealizations about agents' knowledge, the idealization of prescience is intended to remove ignorance that prevents the realization of an objective solution and makes subjective solutions diverge from objective solutions. We can use our account of solutions to make the idealization of prescience more precise. What type of predictive power is needed to make subjective and objective solutions coincide?

The appropriate type of predictive power depends on the nature of objective solutions. We are interested primarily in *relative* objective solutions, profiles where agents are jointly successful. They coincide with relative subjective solutions in ideal games, and so are profiles of jointly rational strategies. However, we also have a side interest in *nonrelative* objective solutions, profiles where agents are individually successful. These are profiles of rational strategies in ideal games. We ask, what idealizations about knowledge are needed in ideal games to make profiles of jointly rational strategies jointly successful, and profiles of rational strategies individually successful?

37

Since strategic reasoning concerns hypothetical situations, and since successful strategic reasoning requires accuracy about those situations, the idealizations entail correct beliefs *about* hypothetical situations, and not just correct beliefs *in* hypothetical situations. In the actual situation an ideal agent knows for every hypothetical situation the information required for successful strategic reasoning. The individual success of strategic rationality requires the accuracy of an agent's prediction of other agents' responses to his strategies. So the appropriate idealization for nonrelative objective solutions is each agent's foreknowledge of the other agents' response to each of his strategies. (We assume that there is a response even if the strategy is counterfactual.)

The appropriate idealization for relative objective solutions is foreknowledge of the profile realized. In ideal conditions a relative objective solution is a relative subjective solution, that is, a profile of strategies that are jointly rational, or rational given the profile. To make relative objective and subjective solutions coincide, agents must have foreknowledge of the profile realized. This foreknowledge is necessary so that agents achieve a relative objective solution if possible. Without it, they may act with different profiles in view and frustrate joint success. A relative objective solution calls for coordination of beliefs. Each agent needs to know the profile realized so that agents respond to the same beliefs about the profile realized. Thanks to foreknowledge of the profile realized, whatever profile it is, a relative subjective solution is not overturned by knowledge of its realization. It is stable with respect to its revelation. This stability with respect to knowledge is necessary for agreement between relative subjective and objective solutions, since it is a basic feature of objective solutions that they are not undermined by general knowledge of their realization.[5]

The predictive power of an ideal agent thus has two varieties according to the type of objective solution of interest: (1) the power to predict the response of other agents to any strategy of his and (2) the power to predict the profile realized whatever it is. These are different predictive powers because an agent's strategy typically appears in several profiles. An agent might be able to predict the response of other agents to a strategy of his. But suppose a profile is realized that contains that strategy and a different response by other agents. Does he predict it? This requires the second type of predictive power, the power to predict whatever profile is realized.

Prediction of profiles does not conflict with prediction of responses to strategies. First, consider the actual outcome of a game. Only one profile

---

[5] An agent's foreknowledge of the profile realized entails his foreknowledge of his own part in the profile. Such foreknowledge does not rule out choice. The agent's foreknowledge of his strategy may actually play a role in deliberations culminating in adoption of that strategy.

is realized. An agent may predict the profile and also predict the strategies of other agents in the profile as their response to his part in the profile. Next, consider counterfactual outcomes of the game. These are counterfactual profiles with at least some counterfactual strategies. The response to a counterfactual strategy is a subprofile. The subprofile may be predicted even if, given the realization of a profile containing the strategy and a different subprofile, that subprofile is also predicted. Supposition of a counterfactual strategy and supposition of a counterfactual profile are different suppositions even if the profile contains the strategy. So the consequences of these suppositions may be different. The first supposition may lead to prediction of one profile although the second supposition leads to prediction of another profile. Not only are the two types of predictive power compatible, the power to predict profiles in fact entails the power to predict responses to strategies, since a strategy and the response to it constitute a profile.

Although we focus on relative solutions, we sometimes treat nonrelative solutions. Also, we sometimes consider the relationship between relative and nonrelative solutions. For expository convenience, we want ideal agents to be poised to achieve both types of solution. So we conflate the distinction between the idealizations concerning predictive power for the two types of solution. We define prescience to include both types of predictive power whatever type of solution we discuss. We adopt the idealization for relative solutions, which entails the idealization for nonrelative solutions. This makes the idealization of prescience stronger than necessary in discussions of nonrelative solutions. But the extra strength is not a problem for the issues we address, and it allows us to keep idealizations constant as we shuttle between relative and nonrelative solutions.

Notice that prescience may exist in games with mixed strategies if it is applied to mixed rather than pure strategies. We shall understand prescience this way. In the mixed extension of a game it implies foreknowledge of the mixed strategies adopted, that is, the strategies of the mixed extension that are adopted, but not foreknowledge of the pure strategies adopted. However, most of our examples are games in pure strategies where prescience does imply foreknowledge of the pure strategies adopted.

Prescience does not conflict with the causal independence of agents' choices in normal-form games. Prescience entails conditionals connecting the strategies of an agent to the strategies of other agents. But the conditionals are in the indicative mood rather than the subjunctive mood and indicate evidential connections rather than causal connections between agents' choices.

Also, prescience does not make strategic reasoning superfluous. An agent normally uses strategic reasoning to calculate responses to his strategies.

When adopting the idealization of prescience, we do not assume that foreknowledge of responses is direct and not a product of strategic reasoning. The idealization of prescience entails foreknowledge of responses but leaves open the method of acquiring that foreknowledge. Furthermore, even if information about responses is direct, there is a role for strategic reasoning. Once an agent knows responses to his strategies, he calculates his incentives to switch strategy given a profile. Then using knowledge of these incentives, he decides upon a strategy. The agent uses strategic reasoning to go from information about his incentives to a rational decision. His decision is a product of strategic reasoning even if his information about responses is direct and is not itself a product of strategic reasoning.

In normal-form games ideal agents have indirect rather than direct foreknowledge of the strategies of other agents since agents' choices are causally independent. One way for an agent's foreknowledge of the strategies of other agents to be *indirect* is for it to be derived via strategic reasoning from the agent's foreknowledge of his own choice. In Matching Pennies suppose, for example, that Row, the matcher, adopts H. Then by strategic reasoning he may infer that Column pursues her incentive to adopt T, so that the profile realized is (H, T). Or he might infer that Column does not pursue her incentive to adopt T and adopts H so that (H, H) is realized. We say that an agent's indirect foreknowledge of other agents' strategies is *strategic* if it is derived using strategic reasoning – that is, using his own strategy as an essential premiss – and *nonstrategic* if it is derived some other way, say, by replication of the reasoning of other agents without appeal to a premiss about his own strategy. Strategic foreknowledge is less robust than nonstrategic foreknowledge. Its content varies with counterfactual suppositions about the profile realized. By and large, an agent's foreknowledge of responses and profiles is strategic in an ideal game of strategy, but we suppose that agents have a type of foreknowledge that makes the idealizations consistent.

Explaining prescience in terms of the strategic reasoning ability of agents, and their information about their game and each other, is an important and challenging task. An explanation must appeal to basic features of ideal agents such as their rationality, the transparency of their incentives, their ability to reproduce the strategic reasoning of other agents, and perhaps even deeper-going features. For instance, ideal agents' insight about each other is often taken to involve common knowledge of certain facts, where *common knowledge* that P is each agent's knowledge that P, knowledge that all know that P, knowledge that all know that all know that P, and so on. Common knowledge of the rationality of agents removes the following sort of predictive failure: He knows she is rational, but she may not know that he is rational, so she may fail to predict him, and not knowing what she expects

40

of him, he cannot predict her. Although I maintain that an explanation of prescience in ideal games is possible, I do not attempt to provide one. I assume prescience rather than derive it from basic features of ideal agents. I assume prescience because its explanation raises issues that concern the realization rather than the existence of equilibria. Our topic is the existence of equilibria, and what may happen without loss of joint self-support, rather than the realization of equilibria, and what does happen. We focus on the nature of equilibria, and whether they exist, rather than on how they are realized, and under what conditions they are realized.[6]

The idealization of prescience is a component of our complete idealization for games. The idealization about rationality is another component. Do these idealizations cohere? In particular, how do we interpret the idealizations of rationality and prescience with respect to counterfactual profiles? We give prescience priority over rationality in counterfactual situations where they conflict, so we allow for counterfactual irrationality. If a nonsolution is counterfactually supposed, then some agent is irrational under its supposition. We take such irrationality to be restricted to the minimum required by the counterfactual supposition. Also, we give priority to the assumption of predictive power for profiles over constancy of predicted response. Predictions of responses are therefore dependent on the profile realized. When a profile is supposed that does not pair an agent's strategy with the response he predicts for it, we have to suppose a new pattern of reasoning for the agent, one that yields a correct prediction of the response to his strategy. To avoid inconsistency between predictive powers for profiles and for responses, we give priority to the assumption of predictive power for profiles over constancy of predictions of responses to strategies.

To illustrate the compatibility of the idealizations about rationality and prescience, take an ideal version of Matching Pennies in which the outcome is (H, H). Row's knowledge of Column's strategy may be nonstrategic and robust. It may stem from his knowledge of Column's psychology rather than from strategic reasoning. Row may have nonstrategic foreknowledge of Column's choice of H as a result of replicating her reasoning. If (H, H) is realized, Row may know that if he switches to T, Column still adopts H. On the other hand, Column's foreknowledge of Row's strategy may be strategic and nonrobust. Suppose Column obtains foreknowledge of Row's choice by reasoning strategically that he picks H in response to her choosing H. Such foreknowledge is not robust and changes with changes in supposition about Column's strategy. Given her strategic foreknowledge of Row's choice, she supposes that Row chooses T if she chooses T. Pursuit of her incentive to

---

[6] An explanation of prescience must overcome formidable difficulties. For example, it must show how without communication rational agents in a normal-form game are able to predict the equilibrium realized when there are several equilibria.

switch from H to T is therefore futile. As a result, Column's choice of H is not necessarily irrational. Granting this point, Column's strategic inference of Row's response to her strategy, and her subsequent failure to pursue her incentive to switch from H to T, may occur in an ideal version of Matching Pennies where the agents are rational as well as prescient.

Besides the idealizations about rationality and prescience, another component of the complete idealization for relative solutions is informedness, that is, knowledge of the game, in particular, the payoff matrix in the case of a normal-form game. What is the relationship between the idealization of prescience and the idealization of informedness? In this case there is no suspicion of incompatibility, but in fact the opposite. Perhaps prescience is derivable from informedness given the other idealizations.

Both the idealizations of prescience and informedness are related to an agent's knowledge. Foreknowledge of strategy profiles and knowledge of the payoff matrix together yield foreknowledge of payoff profiles in normal-form games. Their combination is the idealization of foreknowledge of the outcome of the game in all eventualities. It is useful to split foreknowledge of outcomes into components concerning strategy profiles and the payoff matrix. But prescience and informedness do more than break down into components the idealization of foreknowledge of outcomes. Informedness includes information about the agents' pursuit of incentives. It includes knowledge of incentives pursued, since such knowledge is necessary for successful strategic reasoning. Strategic reasoning has the form, If I do this in response to their doing that, then they respond in such and such a way, and I have such and such incentives to switch given their response. The idealizations for strategic reasoning keep pace with the demands of strategic reasoning, and so provide all information required for effective strategic reasoning. This includes foreknowledge of outcomes. But it also includes foreknowledge of responses to strategy switches, or pursuit of incentives by other agents. This foreknowledge concerns not just the profile realized, but the profile realized if an agent switches strategies. This additional information about agents is included in informedness, or knowledge of the game. In general, the idealization of informedness furnishes all information about the game necessary for successful strategic reasoning.

Although informedness differs from prescience, some foreknowledge involved in prescience may be derivable from the idealizations concerning rationality and informedness, so the idealization of prescience may not be completely independent of other idealizations. However, we do not try to break down the overall idealization for relative solutions into independent component idealizations. For simplicity, we just assume prescience and do not explore the possibility of deriving it partially or wholly from other idealizations.

2.4. EXISTENCE OF SOLUTIONS

Chapter 3's argument against the standard of Nash equilibrium for solutions assumes that every ideal game has a solution (not necessarily a unique solution). This section lays the groundwork for that assumption.

Can a game lack a solution? Can it lack a profile of strategies that are rational given the profile? For a profile to be ruled out as a solution, there has to be an agent whose rational choices given the profile do not include his strategy in the profile. If for every profile there is such an agent, then there is no solution. Can this happen? Our view is that in every decision situation an agent has a rational strategy. Chapter 4 supports this view by showing that a necessary condition of rationality, self-support, is possible in every decision situation. It follows from our view that in every game each agent has a rational choice given the realization of any profile. Does this guarantee the existence of a solution?

A solution is defined as a profile of strategies that are rational given the realization of the profile, not strategies that are nonconditionally rational. The claim that every game has a solution therefore goes beyond the claim that every game has a profile of rational choices. It asserts that every game has a profile of strategies that are rational given the profile. There is a difference between each agent in a game having a rational choice and the agents having choices that are jointly rational. A profile of rational strategies is not the same as a profile of jointly rational strategies. It is an open question whether each agent's having a rational choice entails that some profile consists of choices that are jointly rational.

Can we derive the existence of a solution from the existence of rational choices even though a strategy's belonging to a solution is not the same as its being a rational choice? Let us consider an attempt to do this for normal-form games. The argument tries to move from the causal independence of the choices of agents in a normal-form game to the existence of a solution. It goes as follows:

In every decision situation an agent has a rational choice. So in every game there is a profile of rational strategies. A profile comprising rational strategies is a solution unless the supposition that the profile is realized affects the status of the strategies. In a normal-form game choices are independent. In particular, an agent's information is not affected by the strategies of other agents. So if an agent's strategy is rational, it remains rational when combined with any strategies of other agents. Even if an agent's strategy is rational only because of false beliefs about the strategies adopted by other agents, his strategy remains rational whatever strategies of other agents are supposed. A profile of rational strategies is therefore a profile of strategies rational given the profile, that is, a solution.

43

This argument has a flaw. In some normal-form games the supposition of a profile does carry new information for an agent. This happens, for instance, in ideal games, where the agents are prescient. The supposition that a profile is realized entails the supposition that the agents anticipate its realization. The realization of the profile has no causal influence on an agent's information about the strategies of other agents, but its supposition may influence what is taken as the agent's information about the strategies of other agents. The supposition may indicate new information for the agents about their strategies. As a result, a profile of rational strategies need not be a profile of strategies rational given the realization of the profile. It need not be a profile of jointly rational strategies and so need not be a solution.

To see how this is possible in principle, imagine a two-person game in which the agents are not both rational. Whatever the agents do, rationality requires one of them to act differently. That is, if one agent acts rationally, the other agent does not. Perhaps one and the same agent is irrational whatever strategy he realizes – in every profile, that agent's strategy is irrational given the profile. Or perhaps every profile contains a strategy that is irrational for one agent or another given the profile, a different agent in different profiles. Either way, the game lacks a solution. Whatever profile is realized, some agent's strategy in the profile is irrational. No profile is such that all strategies in it are rational given the realization of the profile.

This situation would occur in Matching Pennies, given prescient agents, if having an incentive to switch strategies were a sufficient reason to switch. For instance, if (H, H) is realized, then Row's choice of Heads is rational, since he wants a match. But Column's choice of Heads is not rational under the assumption about sufficient reasons, since she wants a mismatch. She has a rational choice, namely, Tails. However, if she chooses Tails, then Row's choice of Heads, given his prescience, is not rational under the assumption about sufficient reasons since he has an incentive to switch strategies.

Even putting aside incentive-proofness as a standard of rationality, the rational strategies of agents are still incompatible in some two-person games. This cannot be demonstrated conclusively until we have introduced an interpretation of self-support to replace incentive-proofness. But incompatibility of the rational strategies of agents seems possible since the realization of a profile changes the circumstances with respect to which strategies are assessed. A strategy *rational given a profile* is rational in light of the information the profile carries. Even though an agent has a rational choice in every choice situation, it seems possible that a profile of rational choices is not a profile of jointly rational choices. Some agent's choice, although rational, may not be rational given the profile. That is, some agent's rational choice may not be rational given the rational choices of other agents. In this case the profile of rational choices is not a solution. Section 7.2 presents a nonideal game of this type without an equilibrium and therefore without a solution.

Given that some games lack a solution, we cannot argue against the standard of Nash equilibrium for solutions simply by pointing out that some games lack a Nash equilibrium. Being a Nash equilibrium might still be a necessary condition for being a solution if the games without Nash equilibria are also without solutions. Hence we appeal to the premiss that every *ideal* game (with a finite number of agents) has a solution. Given this premiss, and the additional premiss that some ideal games (with a finite number of agents) lack Nash equilibria, we conclude that being a Nash equilibrium is not necessary for being a solution. Chapter 3 presents this argument in detail.

Let us try to bolster the assumption that every ideal game has a solution. The idealization of rationality we adopt – that agents are jointly rational if possible – forces us to forgo an argument we might have used to support the assumption. The argument does not require a precise characterization of rational strategies. It relies instead on the ideality of agents. It claims that any profile realized by ideal agents is a solution – there is some profile realized by the agents, and that profile is a solution to the game. More specifically, the argument goes as follows:

In ideal games agents are prescient and so anticipate the profile realized. The agents are also rational. That is, all adopt rational strategies so that the profile realized is composed of rational strategies. It then follows that the strategies are rational given the profile realized. Therefore the strategies the agents adopt are jointly rational and form a solution.

Unfortunately, this simple argument assumes an idealization stronger than what we are prepared to allow ourselves. It assumes that in an ideal game all agents are rational. This assumption is too strong from our point of view. We authorize only the idealization that all agents are rational *if it is possible for all to be rational jointly*. To support the assumption of universal rationality using this idealization, we need a supplementary argument that in an ideal game it is possible for all agents to be rational jointly, in effect an argument that an ideal game has a solution. So from our point of view, the foregoing argument does not establish the existence of solutions to ideal games.

Our adoption of the modest idealization that agents are jointly rational if possible also forces us to qualify the claim that every ideal game has a solution. Strictly speaking we claim the existence of solutions only in ideal games *with a finite number of agents*. The restriction is necessary to prevent begging questions. To see this, consider the following:

A game with an infinite number of agents A1, A2, ... lacks a solution if it has profiles P0, P1, P2, ... with the following property: If profile P0 is realized, then agent A1 has a sufficient incentive to switch to profile P1; if P1 is realized, then agent A2 has a sufficient incentive to switch to profile P2; and so on. There is no profile where all agents are rational given the profile. The incentives of agents prevent joint rationality. An ideal version of this game is an ideal game without a solution.

In the face of such doubts, we might try to maintain the intuition that every ideal game has a solution by arguing that games with the features described have no ideal versions. We might argue that there are no ideal games with the features specified instead of conceding that certain ideal games with those features lack solutions. Since given prescience and informedness, it is impossible for all agents in the games to be rational, we might conclude that it is impossible for all agents to be ideal.

This line of reply is not open to us. Our idealization concerning the rationality of agents requires only that agents be jointly rational *if possible*. So the idealizations we advance can be met in such games. We must allow that games with an infinite number of agents may have structural features that prevent solutions even though the games violate none of the idealizations we advance. Hence when we claim that every ideal game has a solution, we implicitly impose a restriction to games with a finite number of agents.

The restriction can be dispensed with if instead of the idealization that agents achieve joint rationality where it is possible we can justify the stronger idealization that all agents are rational. However, we do not try to justify the stronger idealization, since we want our idealizations to leave open the question of the existence of solutions. Given the restriction to games with a finite number of agents, we can get by with the idealization that agents are jointly rational *if possible*. The non–question-begging idealization that agents are jointly rational if possible is strong enough, I believe, to ensure the existence of solutions in games with a finite number of agents. To avoid the controversial, stronger idealizations that are necessary to ensure the existence of solutions in games with an infinite number of agents, we simply restrict ourselves to games with a finite number of agents. We do not deny that ideal games with an infinite number of agents have solutions, but to avoid controversy, we refrain from claiming that they do. We simply claim that ideal games with a finite number of agents have solutions and support this restricted claim using our weak but non–question-begging idealization about joint rationality.

I do not demonstrate that every ideal game has a solution. A demonstration requires a detailed, precise characterization of a solution. According to the definition we have adopted, a solution is a profile of strategies that are rational given that the profile is realized. This characterization of a solution is not detailed and precise enough for an existence proof. For an existence proof, we need a specification of the features a profile must have to comprise strategies that are rational given the profile's realization. Such a characterization of a solution is beyond our ambitions. Our goal is to give a detailed, precise characterization of a necessary condition for being a solution, namely, being an equilibrium. So to support the intuition that every ideal game has a solution, we mainly describe idealizations and point out the way in which they remove obstacles to the existence of solutions. But also in Chapters 4 and 5, respectively, we adopt a new characterization of self-support and a revised standard of equilibrium. Then in Section 5.3 we demonstrate that an equilibrium in the revised sense exists in every ideal game (with a finite number of agents). The proof shows that every ideal game has a profile that meets a necessary condition for being a solution. This lends additional support to the intuition that every ideal game has a solution.

3

Equilibrium

The main tenet of decision theory is that a rational decision maximizes utility or, where there is uncertainty, maximizes expected utility. The main tenet of game theory is that a solution is a Nash equilibrium. The unification of decision theory and game theory requires modifications in the main tenet of each theory. Decisions in games raise complications that call for amendments to decision principles. The amended decision principles in turn call for amendments to solution standards. My project is to revise the main tenets of decision theory and game theory in order to accommodate the existence of an equilibrium in every ideal game. Our revision of the equilibrium standard for solutions rests on revisions of decision principles, in particular, the principle of self-support, a principle that plays an important role in strategic reasoning. This chapter explains the relationships among strategic reasoning, self-supporting decisions, and equilibrium.

3.1. EQUILIBRIUM DEFINED

Solutions and equilibria are closely related. A profile is a solution only if it is an equilibrium. Solutions involve joint rationality, whereas equilibria involve joint self-support, a component of joint rationality. The close relationship between solutions and equilibria guides our definition of an equilibrium. Acknowledging that being an equilibrium is necessary for being a solution, we make the definition of an equilibrium analogous to the definition of a solution:

An *equilibrium* is a profile of strategies that are self-supporting *given that the profile is realized.*

The supposition is made in the same way as for profiles that are solutions.

Our definition of equilibrium uses joint self-support, or self-support relative to a profile. A strategy's being self-supporting given a profile is its being self-supporting given a profile containing the strategy. When we want to emphasize the condition, we say that the strategy's self-support is *relative* to supposition of a profile, in contrast with *nonrelative* self-support, which obtains with respect to actual circumstances only. When we want to emphasize that we define equilibria in terms of relative self-support, we call them *relative* equilibria.

Relative equilibria, or profiles of jointly self-supporting strategies, differ from *nonrelative* equilibria, or profiles of strategies that are self-supporting *tout court*. A relative equilibrium is a profile of strategies that are self-supporting given that the profile is realized, whereas a nonrelative equilibrium is a profile of strategies that are self-supporting, period, that is, self-supporting given the profile actually realized. Relative equilibria depend on responses to a strategy and incentives to switch strategy that are relative to the supposition that a particular profile is realized. Nonrelative equilibria depend on responses to a strategy and incentives to switch strategy given actual circumstances. A relative equilibrium is a profile such that no agent has a sufficient reason to switch from his strategy in the profile given the realization of the profile. In contrast, a nonrelative equilibrium is a profile such that no agent has a sufficient reason to switch from his strategy in the profile, period.

We focus on relative equilibria since they are necessary for relative solutions, our ultimate interest. Working with relative equilibria also has certain incidental advantages. We want to use equilibria to narrow the field of candidates for solutions to ideal games. For this purpose, relative equilibria serve better than nonrelative equilibria. It is easier to identify relative equilibria than nonrelative equilibria. To find nonrelative equilibria, we must be able to compute solutions. In an ideal n-person game we must ascertain responses to strategies in order to discover incentives to switch strategy and then nonrelative equilibria. But a response to an agent's strategy is a solution to an $(n - 1)$-person game played by the other agents. So we must find solutions to subgames to find nonrelative equilibria. On the other hand, we can avoid calculation of solutions to subgames if we look for relative equilibria. Given our idealizations, we can tell whether a profile is self-supporting given the realization of the profile by considering whether any agent has a sufficient reason to switch strategy given the profile. If no agent has a sufficient reason to switch strategy given the profile, the profile is an equilibrium. We need not investigate whether any agent has a sufficient reason to switch strategy given the solution to the subgame obtained by fixing his strategy. The context for reasons to switch is given by the profile itself and does not have to be obtained by calculating solutions to subgames.

Relative equilibrium depends on incentives *relative* to a profile, and nonrelative equilibrium depends on *nonrelative* incentives. In general, relative incentives are easier to work with than nonrelative incentives. In an ideal game, nonrelative incentives depend on the profile actually realized, the profile all agents know is realized. Nonrelative incentives can be specified only if we know the profile actually realized. We usually are not in a position to specify nonrelative incentives since we are not in a position to specify the profile realized. In ideal games the profile realized depends on the profiles that are solutions. But we do not give a full account of solutions. We only formulate a necessary condition for being a solution, namely, being an equilibrium. A full-scale examination of solutions is beyond this book's scope. And without it, we do not have the resources to examine thoroughly nonrelative incentives, equilibria, and solutions.

Since we investigate equilibria rather than solutions, our idealizations for games need one adjustment. In order to avoid begging the question about the existence of solutions in ideal games, we adopted the idealization that agents realize a solution *if possible*, that is, are jointly rational *if possible*. We also want to avoid begging the question about the existence of equilibria in ideal games. Hence, although we hold that every ideal game has an equilibrium, we assume only the idealization that agents realize an equilibrium *if possible*, that is, adopt jointly self-supporting strategies *if possible*. We take this disposition for joint self-support to be robust and to apply to subsets of agents. Furthermore, to make this idealization for equilibria conform with the idealization for solutions, we adopt a restriction to games where an equilibrium exists if and only if a solution does. We take the restriction to be imposed from this point on, although we do not mention it again.

Self-support, like rationality, is with respect to the strategies available and preferences among them. In a game of strategy, preferences may shift with the information carried by an assumption of a profile. In the games we treat, normal-form games, we assume that preferences between outcomes, and utilities of outcomes, are independent of the profile supposed, so that supposition of a profile does not change an agent's preference ranking of outcomes. As a result, we may regard outcomes as utilities. But we do not assume that preferences between strategies and utilities of strategies are epistemically independent of the profile supposed. So a strategy that is self-supporting given one profile may not be self-supporting given another profile. An agent may fail to have a sufficient incentive to switch from a strategy given one profile, but then have a sufficient incentive to switch from the strategy given another profile. In fact, in ideal normal-form games incentives to switch strategies clearly depend on the profile supposed. In the ideal version of Matching Pennies, where agents are prescient, Row, the matcher, has an incentive to switch from Heads given (H, T) but no incentive to switch from Heads given (H, H). Joint self-support takes account of incentives that shift with the profile realized. How this happens is the topic of the next section.

3.2. STRATEGIC REASONING

Chapter 4 formulates a principle of strategic reasoning – that is, a principle for evaluating strategies whose supposition creates incentives to switch to other strategies. The principle is motivated by decision-theoretic considerations essentially independent of the theory of games. Chapter 5 uses that principle to formulate and support a new equilibrium standard for solutions. The principle of strategic reasoning advanced is a principle of self-support. It says that a rational strategy is self-supporting in a certain sense. It fleshes out the idea that an equilibrium is a profile of jointly self-supporting strategies. This section takes a preliminary look at strategic reasoning and self-support.

3.2.1. Self-support

Questions about self-support arise in contexts where strategies create incentives to switch to other strategies. This happens in normal-form games given standard idealizations about the knowledge of agents. We want to show that agents' incentives are relative to choices in normal-form games and that because of this, principles of self-support are needed to evaluate choices. We want to show that rationality requires special decision principles in normal-form games.

When agents' choices are observed in a series of moves in a multistage game of strategy, standard principles of decision making suffice. Rational agents maximize expected utility in the usual way. An agent knows that an opponent will observe his choice, not merely reason about it, and will use her observation to guide her choice. He can predict her response to his own choice in a straightforward even if complex way. To do this, he reasons about what she *would* do if he *were* to do such and such. But when choices are made in a single-stage game, an agent knows that an opponent's choice rests on her reasoning about his choice, not her observation of it. To predict her choice, he reasons about what she *does* if he *does* such and such. Then he considers what his incentives *are* given her response to his strategy. More sophisticated decision rules are needed, in particular, rules incorporating principles about self-supporting choices.

When agents choose in a single-stage game, an agent's choice may carry information about his current, not just his future, situation. His choice may carry information about other agents' choices beyond the information, if any, it carries about the causal influence of his choice on their choices. This information may alter his incentives. Strategic reasoning must take account of the situation that obtains if he *adopts* a strategy, and not just what would happen if he *were to adopt* the strategy. The information his

51

choice carries about the response of other agents is important in evaluating the choice. Any evaluation of the choice that ignores any of that information is insufficiently informed. We need an evaluation of the choice that rests on all the information carried by the choice itself.

Of course, the assumption that a choice is made does not really change information. Only making the choice does that. But, as I say, the assumption of the choice *carries information* that is relevant to the assessment of the choice. When supposition of a choice carries information about incentives, an ideal agent should take that into account before deciding. Evaluating a choice calls for evaluating it under the supposition that it is made, and this requires taking account of the information its supposition carries. The evaluation of a strategy depends on the agent's body of information given its supposition, that is, after the agent's body of information incorporates its supposition and adjusts, say, by adjusting estimations of the strategy's consequences.

One cognitive goal for choice is support by information at the time of choice. This requires support of the choice by information including the choice itself, and so the choice's self-support. An option has to meet this goal if it is not to be regretted as it is chosen. It is rational to pursue the goal of self-support by anticipating one's information given an option. Although a rational choice rests on information in one's possession during deliberation, and during deliberation one may not have the information one will have given an option's adoption, one may have information about the content of the information one will have at the time of choice if the option is adopted. Rationality requires taking account of that information. Rationality demands looking forward from the time of deliberation to the time of choice.

The role of a principle of self-support and, in general, strategic reasoning is to take account of the information a choice carries. An option is a rational choice for an agent only if it is supported by the information the agent anticipates having if he adopts the option. Assessing the expected utility of a strategy with respect to the assumption that it is chosen is one way of taking account of the information that a choice is anticipated to provide, and the incentives that it generates. This approach leads to the principle of ratification discussed in Section 3.2.2. Our principle of self-support, presented in Chapter 4, is a refinement of the principle of ratification.

Let me make a few remarks about terminology. I formulate principles of self-support in terms of *incentives to switch options*. In a decision problem for an agent, I say that the agent has an incentive to switch from option A to option B if and only if he prefers B to A under the assumption that A is adopted. In short, he has an incentive to switch from A to B just in case if A is adopted, he prefers B. Speaking of incentives to switch options is briefer than speaking of conditional preferences. Also, the terminology evokes a useful analogy. It comes from multistage games of strategy where agents have an opportunity in one stage to reverse moves made in earlier stages. Conditional preferences in decision problems have a role similar to the role of preferences for moves in multistage games. Reasoning about self-support in decision problems, in particular, strategic reasoning in single-stage games, is analogous to strategic reasoning in multistage games.

Although speaking of incentives to switch options in decision problems is brief and suggestive, it is potentially misleading for two reasons: First, there is an objective sense of "incentive" according to which an agent's incentives depend on his situation, including circumstances of which he is ignorant. In taking incentives to be dependent on preferences, we are taking incentives in the subjective sense according to which an agent's incentives depend on his information. We are interested in incentives in the subjective sense because our topic is rational decision. We have to take care not to confuse the objective and the subjective senses of incentive.

Second, a decision is a mental event that once made is inalterable, as is any event in the past. Changing one's mind is just making another decision. Therefore, once an option has been adopted, switching is not possible, and if no option has been adopted, then again switching is not possible. In short, switching is impossible, and strictly speaking there are no incentives to switch. An incentive to switch from A to B is better understood as an incentive to switch *consideration* from A to B, or to switch a tentative decision from A to B, so that what is at stake is entertaining A, or leaning toward A, not adopting A. But this view is still a little misleading since it suggests a temporal process of deliberation, whereas decision may be instantaneous and without prior tentative decisions and the like. Although we speak of incentives to switch in decision problems, we have in mind conditional preferences, no more. The same goes for reasons to switch. A reason to switch from option A to option B is just a reason to prefer B to A under the assumption that A is adopted. In particular, in a single-stage game, an incentive to switch strategy is not an incentive to make a move, but rather a conditional preference – a preference for another strategy given the one adopted. An agent has an incentive to switch from one strategy to another if given that he adopts the first strategy, he prefers the second. I hope that pointing out the ways in which the terminology is misleading will prevent the reader from being misled.

Self-support is absence of a sufficient reason to switch. In other words, given a self-supporting strategy, an agent does not have a sufficient reason to switch. We claim that an agent's strategy must be self-supporting in order to be rational. The requirement that an agent's strategy be self-supporting makes a difference in cases where the agent's incentives vary with the choice supposed. In normal-form games an agent knows that his choice itself carries information about the choices of other agents and hence about his payoff.

Given his choice the agent may have an incentive to switch. The incentives to switch need not rest on certainty about the responses of other agents; they may rest on the subjective probabilities of various possible responses. But in ideal normal-form games agents are prescient, and a choice carries full information about the response of other agents and the relevant consequences of the choice. There is no need to consider incentives to switch that arise from mere probabilities of responses, and so no need to use expected utilities rather than utilities. In general, incentives to switch are with respect to expectations concerning the responses of other agents, but in ideal games the expectations amount to foreknowledge. In order for a choice to be rational in a normal-form game, it must take account of the information it is anticipated to provide. No choice is rational if it is *self-defeating*, that is, anticipated to provide information that yields a sufficient reason to switch to another option. Incentives are one type of reason. Hence no choice is rational if it is anticipated to provide information that generates an incentive to switch to another option, and that incentive is a sufficient reason to switch to the other option.

In normal-form games an agent's choice does not causally influence the other agents' choices by, say, causing a change in their information and incentives; choices may be simultaneous, or agents may be prevented from observing the choices of other agents until all choices have been made. Nonetheless, an agent's choice provides evidence about other agents' choices. In the ideal games we consider the agents know enough about each other to anticipate each other fully. So an agent's choice is evidence about the information of other agents and their response to his choice. Because agents choose independently, if an agent *were* to choose differently, there *would be* no change in other agents' information, and hence no change in their choices. But in a typical case, given prescience, if an agent *does* choose differently, then other agents' information *is* different. Consequently, their incentives *are* different, their choices *are* different, and the agent's own incentives *are* different.

In normal-form games, the main strategic issue is whether given the realization of a profile, any agent has an incentive to switch strategy. Incentives to switch strategy depend on causal, not evidential, consequences of alternative strategies. An agent determines whether he has an incentive to switch his strategy in a profile by considering what *would* happen if he *were* to switch to an alternative strategy. According to the common interpretation of subjunctive conditionals, the agent's switch is accompanied by causally minimal departures from the profile by other agents. The agent considers the causally nearest world in which he switches to the alternative strategy. The context, especially the type of game, influences the nearness of worlds and thus events in the nearest world where the agent switches, in particular, switches by other agents. In a normal-form game causally minimal departures from the profile do not include switches by other agents.

Although incentives to switch strategy focus on causal consequences of alternative strategies, self-support for a strategy given a profile attends to the evidential consequences of the strategy itself. When a strategy is evaluated in the context of a profile, the profile and strategy it contains are indicatively supposed, and their evidential consequences set the stage for incentives to switch strategy. The agent considers the evidentially nearest world in which the profile and strategy are realized and calculates his incentives to switch with respect to it. Subjunctive supposition of alternative strategies is embedded in indicative supposition of the profile containing the strategy.

Incentives for an agent to switch strategy may be taken relative to either a profile or a strategy. Since a strategy occurs in many profiles, only one of which gives the other agents' response to the strategy, the incentives to switch away from a strategy relative to a profile may not be the same as the incentives to switch away from the strategy relative to the strategy alone.

Incentives given a profile are not in general the same as incentives given *knowledge* of the profile. But in ideal games, where agents are prescient, supposition of a profile amounts to supposition of knowledge of the profile. Likewise for supposition of a strategy, although in this case the pertinent circumstance is an ideal agent's self-knowledge rather than his prescience. Hence in ideal normal-form games incentives to switch strategy vary with the profile or strategy supposed. Given the realization of a profile, a prescient agent knows his own strategy and the strategies of other agents. Given that information, he may have a new incentive to switch strategy. Also, given a strategy, a prescient agent knows the other agents' response, and given that information, he may have a new incentive to switch strategy.

3.2.2. *Ratification and Equilibrium*

Because incentives are relative to choices in normal-form games, revised decision principles are needed to determine which choices are rational, and, in particular, self-supporting. Some theorists recognize the intricacy of decisions in normal-form games and revise the expected utility principle for those decisions. The most common revision makes the expected utility of a choice conditional on the choice. The expected utility principle then says to adopt a strategy that maximizes expected utility on the assumption that it is adopted.[1] This is the same as requiring a strategy such that given the

---

[1] I put aside the distinction between the assumption that an action is performed and the assumption that the action is chosen by taking options to be actions that are certain to be performed if chosen.

strategy, there is no incentive to switch. Such an incentive-proof strategy is said to be *ratifiable*. Ratification is one interpretation of self-support. As a principle of self-support, the principle of ratification lays down a necessary condition of rationality.[2] Given that an equilibrium is a profile of strategies that are jointly self-supporting, taking self-support as ratification leads to the view that equilibrium is Nash equilibrium.

The principle to pick a ratifiable option has much to recommend it. An option's ratifiability guarantees that it has maximum expected utility with respect to information adjusted for the option's realization. The principle has the virtue of not ignoring information a choice carries. It heeds comparisons of the expected utility of an option with the expected utilities of other options conditional upon the option's adoption. According to the principle, nonconditional rationality involves conditional maximization of expected utility. In order to be rational *nonconditionally* one must adopt an option that maximizes expected utility *given* the option.

We briefly consider the case for the principle of ratification and examine the principle's affiliation with the standard of Nash equilibrium for solutions. In the end we do not accept the principle of ratification or the standard of Nash equilibrium. However, our principle of self-support is similar to the principle of ratification, and we use our principle of self-support to defend our equilibrium standard in a way similar to the way the principle of ratification is used to defend the standard of Nash equilibrium. So the literature on ratification provides important precedents for our position.

The principle of ratification is often advanced as a principle independent of the principle to maximize expected utility. For example, Sobel (1990: 65) takes ratifiability, in addition to maximization of expected utility, as a necessary condition for rationality. But the case for the principle of ratification is stronger if the principle is taken as an interpretation of the principle to maximize expected utility, an interpretation designed to accommodate the information carried by the realization of an option.

Expected utility is calculated with respect to a probability assignment and so with respect to information that generates the probability assignment. Interpretations of the principle to maximize expected utility differ about the information with respect to which the expected utilities of options are to be computed. One possibility is to use the information one has during deliberations. Another possibility is to use the information one has at the time of decision. Obviously, the difference is important only if relevant information arrives during the transition from deliberation to decision. But this commonly occurs in games. An agent's decision provides evidence

---

[2] See, for example, Jeffrey (1983, Sec. 1.7). For the principle's credentials, alternative formulations, and revisions for special cases, see also Weirich (1985) and (1988), Harper (1986), Rabinowicz (1989), Sobel (1990), and Eells and Harper (1991).

about the other agents' decisions, and that evidence affects his estimation of the outcomes of his options, and thus the expected utilities of his options.

An "at-choice" interpretation of the expected utility principle says that a rational decision is an option that has maximum expected utility on the assumption that it is chosen. Since such an option is ratifiable, the at-choice interpretation of the expected utility principle is equivalent to the principle to adopt a ratifiable option. The principle of ratification is the interpretation of the expected utility principle obtained by computing expected utilities with respect to information one anticipates having at the time of choice. It instructs an agent to use during deliberations information he has about the information he has at the time of choice given various hypotheses about his choice.[3]

The argument for the at-choice interpretation of the expected utility principle is simple and compelling. The objective of rational decision making is a choice that fits one's information at the time the choice is made. Choosing in order to maximize expected utility at the time of choice is the rational way to pursue this objective. If one maximizes expected utility with respect to information one has during deliberations, one's decision fits information one has before one's choice. That is, one's decision fits information that is out of date at the time the decision is made. Only current information can supply reasons that support a decision. So it is rational to decide in light of the information that one anticipates having at the time of choice given each option.[4]

The principle of ratification is especially pertinent in games of strategy, where an agent's decision provides information about the decisions of other agents. Whereas the standard, "prechoice" interpretation of the rule to maximize expected utility ignores this information, the principle of ratification takes it into account by comparing conditional expected utilities that register the information a decision carries.

The principle of ratification, as I interpret it, belongs to causal decision theory – presented, for instance, by Gibbard and Harper (1978) – since the expected utilities involved are computed using probabilities that register only

---

[3] In game theory we can generally neglect the distinction between (1) maximizing expected utility with respect to information at the time of decision and (2) choosing an option that maximizes expected utility on the assumption that it is chosen. In standard game situations an agent knows that the only relevant information that will arrive with his decision is the content of his decision. So he knows that the expected utility an option will have if it is chosen is the same as its expected utility on the assumption that it is chosen. In this case choosing in order to maximize expected utility at the time of choice is the same as choosing an option that maximizes expected utility on the assumption that it is chosen.

[4] A time lag between deliberation and choice would ground an argument for using prechoice expected utilities. But an appeal to such a time lag violates the standard idealization of decision and game theory that thought is instantaneous.

the causal influence of options on states of the world.[5] However, the conditionalization of expected utility takes into account the evidence an option provides for states of the world. This use of evidence without regard for causal influence is associated with evidential decision theory, for example, Jeffrey's theory (1983). As a result, some theorists might assign the principle of ratification to evidential decision theory. Although the goal of the principle of ratification is an option that causes good things to happen, it assesses the causal prospects of an option with respect to all relevant evidence, including evidence the option is anticipated to provide. However the principle is classified, the important point is that expected utilities use probabilities of the causal sort. As a result, the principle avoids conflict with the principle of dominance. For instance, in Newcomb's problem – presented by Nozick (1969) – the dominant option, two-boxing, is the only ratifiable option.

Principles of self-support, such as the principle of ratification, involve the assumption that a strategy is realized. These principles are the key to decision-theoretic support for solutions to games of strategy. They permit noncircular strategic reasoning since reasoning in accord with them requires less information than reasoning in accord with the standard version of the principle to maximize expected utility. Classical game theory implicitly abandons nonrelative expected utility principles for the principle of ratification. The principle of ratification is behind classical arguments for the standard of Nash equilibrium.[6]

However, as the next section shows, ratification is too demanding a standard of self-support. In some decision problems no option is ratifiable. If ratification were the standard of self-support, no option would be self-supporting, and so no option would be rational. If, as I hold, every decision problem has an option that is rational, and rationality requires self-support, ratifiability cannot be the right explication of self-support. Furthermore, if ratification were the standard of self-support, some ideal games would lack

---

[5] Notice that the principle of ratification instructs us to suppose *indicatively* that an option is adopted. The indicative mood gives the assumption of an option an epistemic or evidential, as opposed to a causal, significance. However, the principle does not conflict with causal decision theory's injunction to compute expected utility in a causal fashion, using, say, probabilities of subjunctive conditionals as opposed to conditional probabilities. The expected utilities that arise in applications of the principle of ratification may be computed in a causal fashion even though they are relative to an indicative or evidential supposition of an option.

[6] Von Neumann and Morgenstern (1944: 105, 146–8) implicitly use ratifiability to derive the realization of a Nash equilibrium. They assume, roughly, that in an ideal two-person, zero-sum game an agent can use the strategy he adopts to infer the profile realized. A Nash equilibrium is then a profile in which each strategy maximizes expected utility given the profile, and so maximizes expected utility on the assumption that it is adopted. See, for example, Harper (1988, Sec. 2) and (1991: 267–8) and Weirich (1994) for elaborations of this line of argument. Also, Skyrms (1990b, Sec. 5) and Shin (1991) show that given certain common knowledge assumptions the principle of ratification generates correlated equilibria as defined by Aumann (1987).

profiles of strategies that are jointly self-supporting, and so would lack equilibria. To provide for the existence of equilibria in ideal games, and rational options in decision problems, we need a weaker standard of self-support than ratification, or incentive-proofness.

The principle of self-support Chapter 4 advances uses conditional expected utilities, as the principle of ratification does. Yet it does not say that an option is self-supporting just in case it is ratifiable, that is, creates no incentives to switch. It is weaker than the principle of ratification. It provides for the existence of a self-supporting option in every decision problem and an equilibrium in every ideal game.

3.2.3. Strategic Reasoning in Normal-form Games

In normal-form games the challenge to the principle of incentive-proofness is plain. An agent's anticipation of other agents' choices given his own choice causes his incentives to vary with the choice supposed. The incentives that exist given that one choice is made may not be the same as the incentives that exist given that another choice is made. Because of the variation of incentives with choices, it may happen that whatever is chosen, the agent has an incentive to switch, so that no strategy is incentive-proof. In other words, in some normal-form games the principle of incentive-proofness is unsatisfiable. We need a revised principle of self-support since self-support is necessary for a rational choice and since a rational choice exists in every decision problem, including decision problems in normal-form games.

Support for the principle of incentive-proofness also erodes for the following reason: As a result of the relativity of incentives, it may turn out that (1) if an agent *were* to switch strategies from A to B, he *would* gain, but also (2) if the agent *does* switch from A to B, he *would* gain from switching back. In such a case the incentive to switch from A to B is undermined by the switch itself. It appears that the incentive to switch is not really a sufficient reason to switch, and so does not make A self-defeating. A strategy need not be incentive-proof to avoid self-defeat.

In more detail, incentives to switch are undermined as follows: In an ideal normal-form game an agent knows the payoff if he *were* to switch strategies. He knows because he knows the payoff matrix and is prescient. He knows the response of other agents to his strategy and so knows the profile realized. Since the agent's switching does not causally influence the other agents' choices, he knows that if he were to switch, their choices would not change. So he knows the profile that would be realized if he were to switch, and hence the payoff he would receive. He also knows the payoff if he *does* switch. He knows that if he does switch, other agents, being prescient, know his strategy and, being rational, make a best response to it. Being prescient himself, he knows their response to his new strategy. So he knows the profile realized if he switches and is thus able to compute his payoff if he switches. He is also able to compute the payoffs of alternative strategies if he *were* to switch again. He can do this because, again, he knows that other agents' strategies are causally independent of his strategy. He knows that if he were to switch again, their strategies would not change. As a result of all this, an agent may know that if he *were* to switch strategies from A to B, he *would* gain, but if he *does* switch from A to B, he *would* gain from switching back. For these reasons, pursuing an incentive to switch may undermine the incentive. Not every incentive to switch is a sufficient reason to switch. Section 3.3 provides examples.

Considering the unsatisfiability of the principle of incentive-proofness in some decision problems and the existence of self-undermining incentives to switch, I hold that strategies that create an incentive to switch need not be self-defeating. Not every incentive to switch establishes self-defeat. Non–self-defeat, or self-support, is possible despite an incentive to switch since self-defeat requires a sufficient reason to switch, whereas not every incentive to switch is a sufficient reason to switch. Chapter 4 characterizes incentives to switch that are sufficient reasons to switch – I call these *sufficient incentives* – and afterward interprets self-support, equilibria, and solutions in terms of these incentives.

Chapter 4 also argues that in every decision problem at least one option is self-supporting. If for every option, on the assumption that it is adopted, there is an incentive to switch to another option, then not all the incentives to switch are sufficient reasons to switch. Some option is self-supporting despite incentives to switch. Our method of ensuring the existence of self-supporting options is to weaken the requirements for self-support.

The propagation of self-supporting options exacerbates another problem, however. In some decision problems there are several self-supporting options, and they are not equally rational choices. Although being self-supporting is a necessary condition for being a rational choice, it is not a sufficient condition. How should one choose among the self-supporting options? This problem arises for all interpretations of self-support and is not peculiar to the interpretation introduced in Chapter 4. Since it is a general problem, we set it aside. We put off the problem of choosing among the self-supporting options to another occasion. It is an important issue, but not the issue that motivates our revision of the principle of self-support. Our revision is motivated by the apparent absence of self-supporting options in some cases, and not the overabundance of self-supporting options in other cases. We will be content to demonstrate the existence of self-supporting options in all cases.

3.3. THE CASE AGAINST NASH EQUILIBRIUM

An objective Nash equilibrium is a strategy profile in which each agent's strategy maximizes his payoff given the strategies assigned to the other agents. In ideal games the agents' knowledge justifies the interchangeability of payoff increases and (subjective) incentives; greater payoffs correspond to incentives to switch. In this context an objective Nash equilibrium can be subjectively characterized as a strategy profile such that no agent has an incentive to switch strategies given the strategies assigned to other agents. As mentioned in Chapter 1, we generally assume the subjective characterization.

We focus on the standard of Nash equilibrium for normal-form games rather than the more general standard of incentive-proofness for games of all sorts. The argument for taking equilibrium, or joint self-support, to be Nash equilibrium in normal-form games is that self-support is lack of a sufficient reason to switch, and an incentive to switch is a sufficient reason to switch. The argument's first claim holds by definition. An agent's strategy is self-supporting just in case he has no sufficient reason to switch away from it. A reason to switch away from a strategy assumes the strategy and evaluates it with respect to other strategies under that assumption. Self-support is just a positive evaluation of the strategy under the assumption that it is adopted. So the crucial point in the argument is its second claim, that every incentive to switch is a sufficient reason to switch.

The argument's crucial assumption is plausible. It is, in fact, hard to call an agent's strategy rational if given it he prefers another strategy. It seems irrational for an agent to adopt a strategy if given it he has an incentive to switch to another. But the argument is refuted if we can show that, contrary to appearances, an agent with an incentive to switch strategy may lack a sufficient reason to switch. A good place to start is an ideal version of Matching Pennies, where whatever an agent does, he has an incentive to switch. It is impossible for an agent to adopt a strategy from which he does not have an incentive to switch. Hence it is impossible for him to adopt a rational strategy if lacking a sufficient reason to switch requires lacking an incentive to switch. In order to provide for the possibility of a rational choice, something I regard as mandatory, we must therefore recognize incentives to switch that are not sufficient reasons to switch.

The rest of this section refines the foregoing refutation of the decision-theoretic argument for Nash equilibrium and introduces some criticisms of Nash equilibrium related to absent Nash equilibria and solutions that are not Nash equilibria. The points against Nash equilibrium are made using three examples. The first and second concern absent Nash equilibria and the refutation of the decision-theoretic argument for Nash equilibrium. The

	Heads	Tails
Heads	1, -1	-1, 1
Tails	-1, 1	1, -1

Figure 3.1   Matching Pennies.

third concerns solutions that are not Nash equilibria. After the examples, I summarize the case against Nash equilibrium.

I use some normal-form games to illustrate the problems with incentive-proofness as an interpretation of self-support, and to prompt the search for an interpretation of equilibrium to replace Nash equilibrium. To start, let us consider Matching Pennies more carefully. We take it as a game in pure strategies. As such it lacks an objective Nash equilibrium. Figure 3.1 repeats its payoff matrix. In an ideal version of the game agents predict the profile realized, whatever it is. So no subjective Nash equilibrium exists. For every profile, there is some agent who has an incentive for unilateral departure from the strategy assigned to him. For example, given (H, H), Column, who seeks a mismatch, has an incentive to switch from H to T. Since every ideal game has a solution, the game has a solution despite the absence of a subjective Nash equilibrium. (The solution is dependent on the agents' pattern of pursuit of incentives, as explained in Section 7.1.) Hence being a subjective Nash equilibrium is not a necessary condition for being a solution. Consequently, Nash equilibrium is unsatisfactory as a characterization of equilibrium. For being an equilibrium, taken as a profile of jointly self-supporting strategies, is a necessary condition for being a solution.

Since this objection to Nash equilibrium depends on taking equilibrium as a necessary condition for a solution, let us review the reasons for this position. If given an agent's strategy the agent has a sufficient reason to switch, the strategy is not self-supporting; it is self-defeating. This follows from the definition of self-support. Now assuming that the strategies of an agent are incompatible (as in normal-form representations of games), an agent's strategy is not rational if given it the agent has a sufficient reason to adopt another strategy. Self-support is necessary for rational choice. Since (1) a solution is a profile of jointly rational strategies, (2) an equilibrium is a profile of jointly self-supporting strategies, and (3) self-support is necessary for rational choice, being an equilibrium is necessary for being a solution. The strategies that a solution comprises must therefore be jointly self-supporting; otherwise they are not jointly rational. However, there may be several equilibria, and they need not be equivalent with respect to standards of rationality. So being an equilibrium is not a sufficient condition for being a solution.

Some theorists may defend the standard of Nash equilibrium by questioning our treatment of Matching Pennies. They may argue that ideal games in pure strategies do not exist. Harper (1986: 31–2) claims that ideal agents can always generate mixed strategies for themselves. Given mixed strategies, Matching Pennies has a Nash equilibrium, and the problem of missing Nash equilibria does not arise. But we accept ideal games in pure strategies. We do not suppose that all ideal agents have built-in randomizers. Furthermore, in our example we do not have to deny the existence of mixed strategies. It is enough if the salient strategies are pure rather than mixed because, say, some penalty is attached to use of a mixed strategy. The representation of a game does not purport to list all strategies, just the salient ones. An ideal version of Matching Pennies in pure strategies is by definition a game that is adequately represented by its pure strategies, and so has only pure strategies as salient strategies even if it has nonsalient mixed strategies. Given this gloss, an ideal version of Matching Pennies in pure strategies is possible and spells trouble for the standard of Nash equilibrium.

Let me make one other comment about our ideal version of Matching Pennies. It cannot be that Row knows Column responds to incentives to switch and that Column knows Row responds to incentives to switch. Whatever profile is realized, some agent does not respond to incentives to switch. To make the assumptions about Row and Column consistent, we specify that when a profile is realized, Row is the one who does not pursue his incentive to switch, and this is known by all. We still regard Row as rational and the game as ideal, although some would disagree given Row's failure to pursue incentives. We take up this issue in the next chapter and defend the rationality of nonpursuit of incentives that are futile to pursue.

We obtain another objection to the standard of Nash equilibrium – a refutation of its decision-theoretic support – if we consider individual choice rather than profiles of choices. In normal-form games without a Nash equilibrium, it is possible that some agent has no incentive-proof strategy. Whether an agent lacks an incentive-proof strategy depends on the other agents' responses to his strategies. It is possible that whatever profile is realized, he has an incentive to switch. In our example Row cannot meet the standard of incentive-proofness. Whatever strategy he adopts, he has an incentive to switch. For instance, suppose that Row adopts Heads. Then he knows that having predicted his choice, Column responds with Tails. So Row has an incentive to switch to Tails. The situation is similar if Row adopts Tails. Then he has an incentive to switch to Heads. It is impossible for Row to meet the standard of incentive-proofness. Whatever he does, he has an incentive to switch to another strategy. Not only is there no profile of jointly incentive-proof strategies, but also there is no incentive-proof strategy for Row. According to the standard of incentive-proofness for decisions, Row has no choice that is rational. Therefore, the standard must be weakened. If there are any games, in fact, any decision problems, in which some agent lacks an incentive-proof strategy, the standard is too strict. We need a weaker standard of self-support than the standard of incentive-proofness so that in every decision problem there is a self-supporting strategy. The revised, weaker standard of self-support leads to a standard of equilibrium weaker than the standard of Nash equilibrium. Nash equilibrium loses its decision-theoretic support once the decision principle of incentive-proofness is rescinded.

We have just argued against the standard of incentive-proofness for rational choice by identifying an agent in an ideal version of Matching Pennies who cannot meet the standard. But if we are willing to use the intuition that every ideal game has a solution, there is another argument against the decision standard that does not require the identification of a particular agent who cannot meet the standard. The incorrectness of the decision standard is shown by the existence of an ideal game with a solution but without a Nash equilibrium. In an ideal game without a Nash equilibrium, every profile leaves some agent with an incentive to switch. Since there is a solution, some profile has strategies that are jointly rational even though not all are incentive-proof. Some profile has strategies that are jointly rational even though some agent has an incentive to switch strategies. Incentive-proofness therefore is not necessary for joint rationality and so is not necessary for rational choice.

The principle of incentive-proofness requires an agent to pursue incentives relentlessly. In some cases, such as an ideal version of Matching Pennies, relentless pursuit of incentives produces a cycle. In our ideal version of Matching Pennies, Row's relentless pursuit of incentives produces the following cycle of strategies: H, T, H, ... Such cyclical pursuit of incentives is futile and so irrational. Cyclical pursuit of incentives is psychologically possible but rationally forbidden.

In general, the equilibrium standard of incentive-proofness rests on the decision principle of incentive-proofness or ratification. This principle says to maximize expected utility relative to the option adopted. But the decision principle is mistaken. Not every incentive to switch is a sufficient reason to switch. The principle of incentive-proofness ignores the futility of pursuing some incentives to switch. Since the decision principle of incentive-proofness is incorrect, the equilibrium standard of incentive-proofness is without decision-theoretic support.

Our second example raises the same two objections to Nash equilibrium. It is an ideal game in which each of two agents picks a positive integer, the

	1	2	3	...
1	0, 0	0, 1	0, 1	...
2	1, 0	0, 0	0, 1	...
3	1, 0	1, 0	0, 0	...
.	.	.	.	
.	.	.	.	
.	.	.	.	

Figure 3.2   The Higher Number Game.

agent with the higher number wins a dollar, and neither agent wins in case of a tie. The agents have an infinite number of pure strategies. See Figure 3.2.

It is impossible for both agents to pursue incentives. So suppose that Column pursues incentives to switch and that both agents know this. Chapter 4 argues that Row is rational despite his failure to pursue incentives. In this game, for each strategy profile, at least one agent has an incentive to switch strategies. So there is no Nash equilibrium. For example, given the strategy profile (1, 2), Row knows Column's choice and so has an incentive to switch from 1 to 3. If being a Nash equilibrium were a necessary condition for being a solution, there would be no solution, contrary to the intuition that every ideal game has a solution. We want to define a type of equilibrium that exists in games of this sort and so can serve as a necessary condition for solutions.

Notice that in the Higher Number Game, the introduction of mixed strategies does not generate a Nash equilibrium. For every profile of probability distributions over pure strategies, some agent has an incentive to shift probability to higher numbers. The mixed extension of a game where agents have an infinite number of pure strategies does not necessarily have a Nash equilibrium. The problem of absent Nash equilibria cannot invariably be dodged by moving to a game's mixed extension.

Again there are difficulties deeper than the absence of Nash equilibria. The game also casts doubt on the decision-theoretic support for Nash equilibrium. Whatever number Row chooses, he has an incentive to switch. If Row chooses 100, he predicts a larger number chosen by Column, say 101, and so has an incentive to switch to a still larger number. Row cannot meet the decision standard of incentive-proofness. Such decision problems demonstrate the need for a revision of the standard of incentive-proofness for decisions, and then the standard of Nash equilibrium for games.

Ideal games without Nash equilibria and decision problems in which an agent has no incentive-proof strategy sufficiently motivate the search for a

	C 1	C 2	C 3
R 1	1, 1	1, 1	1, 1
R 2	1, 1	6, 5	5, 6
R 3	1, 1	5, 6	6, 5

Figure 3.3   An unattractive Nash equilibrium.

standard of equilibrium weaker than the standard of Nash equilibrium. But let me also mention one other motivation – the problem of unattractive Nash equilibria, or apparent solutions that are not Nash equilibria. Take the ideal normal-form game in pure strategies depicted in Figure 3.3. (R1, C1) is the unique Nash equilibrium. If agents adopt strategies from which there is no incentive to switch, they realize that profile. But profiles involving R2 and R3, and C2 and C3 are more attractive. In fact, R2 and R3 dominate R1, and C2 and C3 dominate C1. Intuitively, (R1, C1) is not a solution. So it appears that being a Nash equilibrium is not necessary for being a solution, contrary to the standard of Nash equilibrium. In using this example to criticize the standard of Nash equilibrium, however, I need not lean on the intuition that (R1, C1) is not a solution. I need only maintain that some profile such as (R3, C3) is a solution. A solution that is not a Nash equilibrium is enough to force revision of the standard of Nash equilibrium. We need a weaker equilibrium standard that a profile such as (R3, C3) satisfies.

Let me recapitulate the case against the standard of Nash equilibrium for solutions. There are three reasons to revise the standard. Each argues that Nash equilibrium is not a necessary condition for a solution. First, there are ideal games with solutions but without Nash equilibria. Second, incentive-proofness is not a prerequisite of rational choice. Acknowledging this undermines the argument that Nash equilibrium is a necessary condition for a solution. Third, there are games with unattractive Nash equilibria. Some profiles that are not Nash equilibria are intuitively solutions.

The argument based on unattractive Nash equilibria is straightforward. It rests only on intuitions about solutions. The other two arguments rest on some theoretical principles. They are more powerful arguments since intuitions about solutions are controversial. Let us further clarify those arguments.

Consider the argument based on absent Nash equilibria. It begins by assuming that every ideal game has a solution. Here a solution is taken in the subjective, not the objective sense; a solution is defined in terms of joint rationality. A proof of the assumption requires a precise definition of a rational strategy, something beyond the scope of this book. However, the assumption has strong support. Idealizations remove obstacles to solutions. Since in an ideal game every obstacle to a solution has been removed, it is plausible that a solution exists. Joint rationality is structurally akin to joint morality. Since it is plausible that joint morality is always feasible in ideal conditions, it is plausible that joint rationality is always feasible in ideal conditions.

Next, the argument based on absent Nash equilibria claims that in some ideal games no profile is incentive-proof; every profile if realized has a strategy from which there is an incentive to switch. This happens in ideal normal-form games, such as Matching Pennies, that have no Nash equilibrium. A Nash equilibrium is just an incentive-proof profile. Where there is no Nash equilibrium, the standard of incentive-proofness is impossible for agents to meet jointly. No profile is made up of strategies that are jointly incentive-proof. According to the standard of incentive-proofness, agents cannot adopt jointly rational strategies. Hence there is no solution. This contravenes the intuition that these ideal games have a solution. Since being an equilibrium is a necessary condition for being a solution, Nash equilibrium is not an adequate account of equilibrium. The existence of ideal games with solutions but without incentive-proof profiles forces a weakening of the standard of Nash equilibrium and, in general, incentive-proofness. We need a revised equilibrium standard that can be met in every ideal game.

The other theoretical argument against the standard of Nash equilibrium, the refutation of its decision-theoretic support, dispenses with the controversial assumption that every ideal game has a solution. It uses the less controversial assumption that an agent has a rational choice in every decision problem. This assumption is less controversial since it does not involve highly theoretical concepts, such as the concept of a solution, but only relatively basic concepts, such as the concept of a rational choice. Nonetheless, the assumption effectively undercuts the decision principle that supports the standard of Nash equilibrium. The assumption therefore yields a powerful argument for revising the standard of Nash equilibrium.

The case for the standard of Nash equilibrium rests on the decision principle of incentive-proofness, a particular version of the principle of self-support. But incentive-proofness is not a necessary condition of rational choice. An agent may face a decision problem that lacks an incentive-proof option. In some cases it is impossible for an agent to adopt a strategy such that given it he does not prefer an alternative strategy. We have just seen cases where an agent lacks an incentive-proof strategy, and so must adopt a strategy whose adoption provides incentives for other strategies. We insist on attainable standards of rational choice. Given our assumption that an agent has a rational option in every decision problem, incentive-proofness is not a necessary condition of rational choice. Incentive-proofness is not an appropriate interpretation of self-support. We reject incentive-proofness as a necessary condition of rational choice because it is always possible to make a rational choice, but it is not always possible to make an incentive-proof choice. Rejection of the decision principle of incentive-proofness dismantles the decision-theoretic argument that being a Nash equilibrium is a necessary condition for being a solution. The argument for the standard of Nash equilibrium fails because it assumes an overly strict standard of rational choice.

Once we reject the principle of incentive-proofness for rational choice, we also reject the standard of Nash equilibrium for solutions. The argument for the standard assumes the mistaken decision principle. If lack of a sufficient reason to switch does not require lack of an incentive to switch, then an equilibrium, or profile of jointly self-supporting strategies, is not necessarily a profile of strategies that are incentive-proof taken jointly. In an ideal normal-form game, where payoffs indicate incentives, a profile is an equilibrium if and only if for each agent no strategy has a higher payoff than his part in the profile, given the strategies of others, *unless* the higher payoff is not a sufficient reason for him to switch strategies. The caveat about the higher payoff's being a sufficient reason to switch is necessary. We have to abandon the idea that an incentive to switch is invariably a sufficient reason to switch. We must look for a new characterization of sufficient reasons to switch, and hence self-support, and use it to characterize equilibrium. Revising our principle of self-support alters the standard of equilibrium that the principle grounds.

The problems with incentive-proofness point the way to a revised account of self-support and equilibrium. Reasons to switch from one option to another assume the first option and compare it with the second option under the assumption. Absence of sufficient reasons to switch indicates that an option supports itself. Self-support is the absence of sufficient reasons to switch, or the absence of self-defeat. An equilibrium, a profile of jointly self-supporting strategies, can then be characterized by characterizing sufficient reasons to switch strategy. Traditionally an agent is thought to have a sufficient reason to switch from one strategy to another if given the first strategy he prefers the second, in other words, if he has an incentive to switch from the first strategy to the second. However, we have seen cases that show that the principle of incentive-proofness is too strict. Hence we must revise the view that an incentive to switch invariably constitutes a sufficient reason to switch. Incentive-proofness is too strict an interpretation of the absence of sufficient reasons to switch, and so too strict an interpretation of self-support. Incentive-proofness is not a genuine prerequisite of rational choice. We must recharacterize sufficient reasons to switch strategy in order to respect the existence of rational choices and the necessity of self-support for rational choice.

3.4. DEFENSES OF NASH EQUILIBRIUM

Chapter 4 weakens the characterization of self-support, or, in other words, strengthens the characterization of sufficient reasons to switch, and this leads to a new, weaker equilibrium standard for solutions. The new equilibrium standard accommodates ideal normal-form games without Nash equilibria. Some theorists resist this line of revision and defend the standard of Nash equilibrium. This section replies to some of their points.

The problem of finite normal-form games without Nash equilibria is often put aside by introducing mixed or randomized strategies that generate Nash equilibria. That is, theorists often bypass finite games in pure strategies without Nash equilibria and treat instead their mixed extensions, where the existence of Nash equilibria is guaranteed. They also put aside as unrealistic infinite games that do not have Nash equilibria even granting mixed strategies. This is an unsatisfactory way of addressing the absence of Nash equilibria. We cannot obtain a general standard of equilibrium by ignoring games where Nash equilibria do not exist. A general theory of equilibrium must address games without Nash equilibria, even infinite games without Nash equilibria. Restricting the standard of Nash equilibrium to games where Nash equilibria exist is defeatist. Some other, more general standard of equilibrium is needed for all games, including those in which Nash equilibria do not exist. Furthermore, equilibria of this other more general sort may exist in games where Nash equilibria do exist. So perhaps a solution to a game with a Nash equilibrium is the other sort of equilibrium rather than a Nash equilibrium. This possibility is suggested by games with unattractive Nash equilibria. Nash equilibrium, even where it exists, needs a defense.

Some theorists say that a normal-form game without a Nash equilibrium has no solution – hence an alternative equilibrium standard is not really needed to accommodate it. These theorists dispute the intuition that every ideal game has a solution, or else they deny that normal-form games without Nash equilibria have ideal versions. One argument taking the latter line is suggested by Gibbard (1992: 223). It rejects the possibility of an ideal version of Matching Pennies in pure strategies. It holds that if an ideal version of the game were possible, then both agents – being rational, prescient, and informed about the payoff matrix – would win. Since it is impossible for both agents to win, an ideal version of the game is impossible.

There are two points to make about this argument: The first is that it assumes a stronger idealization than we assume. It assumes that ideal agents are rational, whereas we assume only that ideal agents are jointly rational if possible. We do not dispute the stronger idealization, however, since we adopt the weaker idealization only to avoid begging the question about the existence of solutions in ideal games.

The second point is that the argument assumes that there are no circumstances in which a rational agent fails to pursue an incentive. I have already argued against this assumption by pointing out that it entails that there are decision problems in which an agent has no rational option. Chapter 4 also argues that the futility of pursuit of incentives in some cases allows for a rational failure to pursue an incentive. Section 7.1 applies this lesson to Matching Pennies. Let me briefly explain here why an agent in an ideal version of Matching Pennies may fail to pursue an incentive. The explanation sketches background conditions that permit one of the agents to fail to pursue incentives without being nonideal.

Suppose that nature makes some ideal agents defer to others in games of conflict where one agent must lose. Whenever a deferrer meets a nondeferrer in an ideal version of Matching Pennies, the deferrer fails to pursue incentives and both know that he does. When a deferrer is matched against a nondeferrer, whatever the deferrer does he passes up an incentive to switch and loses. Although if he *were* to switch strategy, he *would* win, if he *switches* strategy, he *does* not win. Despite this, the deferrer is able to pursue incentives. Being a deferrer just means being one who as a matter of fact defers. It does not mean being one who is compelled to defer. It does not mean being unfree to pursue incentives. The deferrer has an incentive to switch strategy since if he *were* to switch, he *would win*. And he is able to pursue the incentive. But he and his opponent know he does not pursue it. Because of his disposition, the deferrer is at a disadvantage in Matching Pennies. It is not his fault. His deference is built in, not the result of on-the-spot deliberation. A deferrer may be an ideal agent. He may be cognitively perfect and fully aware of his circumstances, including his opponent's relevant thought processes. He may also be part of a pair of agents that are jointly rational if possible. Moreover, even though our idealization does not demand it, he may be rational. Rationality does not require that all incentives be pursued, not in cases where it is impossible for all agents to pursue incentives. Nonpursuit of an incentive by the deferrer is rational since pursuit of incentives is futile when he plays Matching Pennies with a nondeferrer. Incentives are just one type of reason for choice; futility of pursuit of incentives is another, contrary reason. When the deferrer is matched against the nondeferrer, he knows that whatever he does he loses, and the futility of pursuing incentives to switch gives him a reason to ignore his incentive to switch. Finally, the deferrer may also be *fully rational*. That is, none of the traits and dispositions he brings to the decision problem need involve any irrationality in him. His disposition to defer is inborn and ineradicable during the period before his choice, and his competition with a nondeferrer may be involuntary.

In concession to proponents of the argument against ideal versions of normal-form games without Nash equilibria, I admit that it is impossible for ideal agents who are psychologically identical and in a perfectly symmetrical situation to play Matching Pennies on the basis of strategic considerations alone. Given parity, there is no reason for the outcome to favor either agent. So if the outcome is determined by strategic considerations alone, it must be symmetrical. Since it cannot be symmetrical, the case cannot arise. Every game in our sense has an outcome, so there is no ideal and symmetrical version of Matching Pennies.

I claim that every payoff matrix for a normal-form game type has an ideal realization, but this claim does not entail that every partial specification of a normal-form game has an ideal realization. Suppose, for instance, that it is specified that both agents in Matching Pennies are deferrers, or that both are nondeferrers. This partial specification of the game has no ideal realization. It is impossible for two deferrers, or for two nondeferrers, to play an ideal version of Matching Pennies in pure strategies. If the agents are ideal, they are aware of the profile realized and their incentives with respect to it. Hence whatever profile is realized, one agent passes up an incentive and the other does not. The game forces exactly one agent to lose, and so requires that exactly one ignore an incentive to switch strategy.

However, ideal games do not require psychological identity and the like. Hence ideal normal-form games without Nash equilibria are possible. Our position is that there is an ideal (asymmetrical) version of Matching Pennies in pure strategies despite the absence of a Nash equilibrium. Ideal versions of normal-form games without Nash equilibria are possible since standards of rationality are accommodating. Rational agents need not meet the standard of incentive-proofness. Games where some agent has no incentive-proof strategy force us to weaken standards of rationality to allow for a rational choice.

The conditions that make a version of Matching Pennies nonideal are psychological conditions of the agents that provide excuses for failing to meet objective standards of rationality. Although Matching Pennies lacks a Nash equilibrium, it nonetheless has an ideal version. It has a version in which neither the agents nor their knowledge of their situation is defective. It has a version in which the agents are jointly rational if possible, informed, and prescient. In such a version of Matching Pennies, there is no extenuating circumstance such that removing it produces a Nash equilibrium. Introducing mixed strategies generates a Nash equilibrium. But the lack of mixed strategies is not an extenuating circumstance. It is not a condition that creates excuses for decisions that fail to meet objective standards. It is just an objective feature of the decision problem, a limitation on the strategies available.

Some theorists deal with the problem of absent Nash equilibria by arguing that being an equilibrium is not a necessary condition for being a solution. Some, such as Kadane and Larkey (1982: 115–16), suggest that (1) a

71

solution is a profile of rational strategies and (2) a profile of rational strategies is a profile of strategies that maximize expected utility. On their view, any profile of expected utility maximizing strategies is a solution, even if it is not an equilibrium. The strategies in a solution do not have to maximize expected utility *given* the solution, so long as they maximize expected utility *tout court*.

Section 1.4 shows that there is good reason to reject the first step in this maneuver. A solution is a profile of *jointly* rational strategies, not just a profile of rational strategies. But let this pass. As far as the connection between equilibria and solutions goes, the second step presents the main claim, the claim that a rational strategy maximizes expected utility nonconditionally. Section 3.2 on strategic reasoning gave good reasons for disagreeing. Maximizing expected utility nonconditionally, without taking account of the information a choice carries about other agents' choices, is not rational. It produces a choice that maximizes expected utility with respect to superseded information. It ignores self-support. An evaluation of a choice should take account of the assumption that the choice is made. Rational choice does not depend on maximizing expected utility with respect to a fixed body of information unadjusted for the strategy assumed to be adopted. Classical game theory is right to take rational choice to require self-support, and being a solution to require being an equilibrium.

Another way to defend the standard of Nash equilibrium is to resist the attack on the decision-theoretic argument for it. That argument for the standard uses the decision principle of incentive-proofness, and criticism of the argument disputes that principle. The case against the decision principle of incentive-proofness assumes that an agent has a rational choice in every decision problem and points out that in some decision problems an agent lacks an incentive-proof option. One can resist the criticism by disowning the assumption that every decision problem has a rational option. Some theorists in fact reject that assumption. They maintain that the concept of rationality breaks down in decision problems where an agent lacks an incentive-proof option. They say that in such decision problems intuitions about rationality are not trustworthy; the decision problems are pathological. But intuitions about rational choice do not break down in these decision problems; the intuitions are merely novel or used in a novel way, as the next chapter shows.[7]

---

[7] Incentive-proofness is another term for ratifiability. Some authors claim that there are no principles of rationality for decision problems without ratifiable options. Jeffrey (1983, Sec. 1.7) says that cases without ratifiable options are pathological. He does not think that Bayesianism can be extended to them. Harper (1986: 33) claims that principles of rational decision break down when there are no ratifiable options. He relies on mixed strategies as a means of obtaining ratifiable options in cases in which no pure strategies are ratifiable. I hold that cases without ratifiable options are difficult, but not untreatable. As Richter (1984: 395–6) points out, we have intuitions about rational decisions in such cases. Principles of rationality can be formulated to generalize those intuitions.

The existence of intuitions about rational choice in games without Nash equilibria grounds the hope that solutions can be discovered and defended. I plan to use intuitions about rational choice to derive solutions rather than use intuitions about solutions to obtain solutions directly. Intuitions about solutions are less reliable than intuitions about rational choice, since the concept of a solution is a construct of a theory of joint rationality for agents in games. In this book I use intuitions about rational choice in games to formulate a necessary condition for solutions; I use them to formulate a new, weaker standard of equilibrium for solutions.

4

Reasons and Incentives

This chapter formulates a principle of rational decision expressing the view that a rational decision is self-supporting. It distinguishes self-support from incentive-proofness and argues that incentive-proofness is not necessary for a rational decision since pursuit of some incentives is futile. The pertinent type of futility is cognitive rather than causal. Pursuit of an incentive is cognitively futile if pursuing it yields no escape from the cognitive state of having an incentive to pursue. This is the case if pursuit of the incentive is part of a pattern of endless pursuit of incentives, a pattern that arises, for instance, if pursuing any incentive undermines the cognitive grounds for the incentive and generates a new incentive.

4.1. FUTILE PURSUIT OF INCENTIVES

Suppose that someone is playing an ideal version of Matching Pennies and is seeking a match. It is impossible for him to randomize his choice by, say, flipping his penny, and he knows that his opponent can anticipate his choice by replicating his reasoning for it. He is considering showing Heads but has the following reason for showing Tails instead: If I show Heads, then my opponent will have anticipated my choice and will show Tails, so my payoff would be higher if I were to show Tails instead. This reason for showing Tails instead of Heads is obviously not sufficient, for there is a similar reason for showing Heads instead of Tails. Having an incentive to switch from one option to another is not in general a sufficient reason for making the switch.

As we learn from an ideal version of Matching Pennies, an incentive to switch strategies is not a sufficient reason to switch. There are reasons besides incentives to switch that bear on the rationality of switching. In particular, there are reasons related to whether the switch is self-defeating, that is, whether the switch, if made, undermines the incentive for it. In an ideal version of Matching Pennies, for instance, it counts against Row's switching from Heads to Tails that if Tails is chosen, Tails is less attractive than Heads. Although it is true that if Row *were* to switch to Tails, he *would* receive a higher payoff, it is also true that if he *does* switch, then it *is* the case that if he *were* to switch back, he *would* receive a higher payoff. The switch to Tails is self-defeating since Row knows that if it is made, the reasons for

74

it disappear. That a switch is self-defeating is a reason against it despite an incentive to make the switch.

The difference between incentives to switch and reasons to switch is especially visible in ideal normal-form games with multiple Nash equilibria. For each Nash equilibrium it is true that given the profile no agent has an incentive to switch from his part in it. Still some considerations provide reasons that may favor one Nash equilibrium over the others. One profile is *Pareto superior* to another if and only if no agent prefers the second and some prefer the first. A profile is *Pareto optimal* in a set of profiles if and only if none in the set is Pareto superior to it. Agents may have reasons to switch to the profile Pareto optimal among Nash equilibria even if they lack incentives to switch to their parts in it. For each agent the switch may be self-supporting because if it is made there is no incentive to reverse it. Making the switch may carry evidence that other agents have adopted their parts in the Pareto optimal Nash equilibrium.

This point about Pareto optimal Nash equilibria is reinforced by keeping in mind that the sort of conditional rationality used to define solutions and relevant reasons is the strong, noncompensatory sort rather than the weaker, compensatory sort. It may be rational in the compensatory sense for an agent to do his part in a nonoptimal equilibrium given that equilibrium, since any irrationality in the equilibrium, except perhaps his own, is excused under this type of conditional rationality. But it may not be rational in the noncompensatory sense for him to do his part in the nonoptimal equilibrium given that equilibrium, since no irrationality in the equilibrium is excused under this type of conditional rationality. The noncompensatory type of conditional rationality entertains reasons against the strategies in a nonoptimal equilibrium that the compensatory type of conditional rationality puts aside. If an agent's strategy in the nonoptimal equilibrium is the reason other agents participate in that equilibrium instead of the optimal one, then the strategy is not rational in the noncompensatory sense, even if it is rational in the compensatory sense. There is a reason against the strategy even if not an incentive to switch from it.

Note that a sufficient reason to switch from an option A to another option B is different from a sufficient reason to adopt B. The reason is sufficient for rejecting A, but not necessarily sufficient for adopting B. Given A, there may be options besides B to which there is a sufficient reason to switch, and one of them may be preferred to B given A, so that it rather than B garners sufficient reasons for adoption. Moreover, reasons to switch option are conditional and change as the option on which they are conditional changes. So even if B is optimal given A, there may be a sufficient reason given B to switch to another option. The reasons for adopting B may then be insufficient because acting on them is self-defeating.

The converse relationship between sufficient reasons for adoption and sufficient reasons for switching is reliable, however. A sufficient reason to switch is a reason that warrants switching. That is, a reason to switch from B to another option is sufficient if and only if in virtue of it B should not be adopted, that is, if and only if it warrants rejecting B. If there is a sufficient reason to adopt B, so that it is a rational choice, then there is no sufficient reason to switch from it to another option. Absence of a sufficient reason to switch from B to another option is a necessary condition for B's being a rational choice, although presence of a sufficient reason to switch to B is not a sufficient condition for B's being a rational choice.

An incentive to switch from option A to option B is a preference for B given A. The foregoing shows that reasons to switch away from a strategy are distinct from incentives to switch, or preferences conditional on the strategy. But furthermore, reasons for a strategy are distinct from incentives to adopt the strategy, or nonconditional preferences for the strategy over other strategies. The latter distinction is apparent in cases where equilibria are weak. Take an ideal version of the mixed extension of Matching Pennies. The unique Nash equilibrium is the profile in which each agent adopts the mixed strategy of playing Heads with probability .5 and Tails with probability .5. Suppose an agent knows that his opponent does his part in this profile. Then for the agent any mixed strategy has the same expected utility as his Nash strategy. No mixed strategy is preferred to another. Nonetheless self-support recommends his Nash strategy. Only his Nash strategy is self-supporting; it is the only strategy that is a top preference on the assumption that it is adopted. So the reasons for the Nash strategy are distinct from the incentives supporting it.

As a final example of the distinction between reasons and incentives, consider the decision problem of picking your own income. Whatever income you choose, you have an incentive to choose a higher income. Whatever option you choose, there is another you prefer. The reasons for choosing an income are therefore different from the incentives for it. The reasons for picking one income instead of a preferred income must include considerations besides incentives. They may address, for example, the futility of relentless pursuit of incentives and the satisfactoriness of incomes beyond a certain level.[1]

Incentives and sufficient reasons are distinct. An incentive to switch options is not the same as a sufficient reason to switch options. Some incentives to switch are not sufficient reasons to switch. In general, the reasons for and against an option comprise more than incentives to adopt options

---

[1] Those worried about practical problems concerning the currency in which income is paid may substitute level of wealth for income, and those worried that level of wealth has a maximum may substitute level of utility in a possible world where that has no maximum.

and incentives to switch options, or preferences and self-conditional preferences. Whether an incentive to switch constitutes a sufficient reason to switch depends on considerations such as the global structure of incentives to switch. The entire incentive structure, or set of preference relations and conditional preference relations among options, is relevant.

Let me clarify a point about reasons and incentives. If an agent has an incentive to switch options, then if he were to switch, his expectations would be higher. Some causal decision theorists say that in consequence every incentive to switch is a sufficient reason to switch. I disagree in cases in which it is impossible to adopt an option such that if it is adopted the agent has no incentive to switch. Taking every incentive to switch as sufficient, there would be a sufficient reason to switch from every option, and no option would be rational, contrary to the intuition that in every decision problem there is a rational choice.

My rejection of the view that every incentive to switch is sufficient does not, however, signal a rejection of causal decision theory in favor of evidential decision theory. Causal decision theory's main contribution is a causal account of incentives. I accept a causal rather than an evidential account of incentives. I do not claim, for instance, that in an ideal version of Matching Pennies Row lacks a sufficient reason to switch from Heads to Tails given (H, T) because he has no incentive to switch. It is true that if he switches to Tails, his prospects with Tails are still bad; his evidence if he switches indicates no improvement. But on my view these evidential considerations do not indicate the absence of an incentive to switch to Tails. The victory that would be won if he were to switch to Tails creates an incentive to switch to Tails. Incentives to switch depend on causal, not evidential, relations. I do not dispute the causal account of incentives to switch, but only the view that every incentive to switch is a sufficient reason to switch.

A solution is a strategy profile – an assignment of strategies to agents – such that given the profile each strategy in the profile is a rational choice for the agent to whom it is assigned. A necessary condition for a profile's being a solution is the absence given the profile of a sufficient reason to switch strategy for any agent. An agent's choice is rational only if self-supporting, and hence only if the agent does not have a sufficient incentive to switch. Since self-support entails the absence of a sufficient incentive to switch strategy, and an equilibrium is a profile of jointly self-supporting strategies, being an equilibrium is a necessary condition for being a solution. To flesh out this important connection between equilibria and solutions, we have to state when an agent has a sufficient reason to switch strategy. We need to specify the incentives to switch that provide sufficient reasons to switch in order to formulate a precise equilibrium standard for solutions to games.

In the course of filling out the equilibrium standard for solutions, puzzles arise about ideal games without a Nash equilibrium because we tend to conflate incentives to switch strategy and reasons to switch strategy. We are tempted to say that these ideal games have no solution, or to renounce the idea that equilibrium is necessary for a solution. We can solve the puzzles, however, if we distinguish reasons and incentives to switch. We can reconcile the intuitions that a solution must be an equilibrium and that every ideal game has a solution. The distinction between incentives and reasons provides a way of defining equilibrium so that being an equilibrium is a necessary condition for being a solution and nonetheless equilibria exist in all ideal games, even ideal versions of Matching Pennies.

An agent's strategy is self-supporting if and only if given it the agent lacks a sufficient reason to switch strategies. If we take every incentive to switch as a sufficient reason to switch, we end up taking equilibrium as subjective Nash equilibrium. We say that an equilibrium is a profile such that given it no agent has an incentive to switch strategies. But once we reject the view that every incentive to switch is a sufficient reason to switch, the standard of Nash equilibrium is not inevitable. According to the basic account of self-support, an equilibrium is a profile such that given it no agent has a sufficient reason to switch strategy. An ideal version of Matching Pennies has an equilibrium according to this basic account if given some profile no agent has a sufficient reason to switch strategy, even though given the profile some agent does have an incentive to switch strategy. Distinguishing reasons to switch and incentives to switch opens up the possibility of alternatives to Nash equilibrium.

Of course, to capitalize on this approach to non-Nash equilibrium, we must say more about sufficient reasons to switch strategy as opposed to incentives to switch strategy. We must provide a new interpretation of self-support by providing a new account of sufficient reasons to switch strategy. This chapter provides a new account of sufficient reasons to switch strategy. The new account describes conditions under which incentives to switch are sufficient reasons to switch, or sufficient incentives. I propose principles of rationality for individual decision-making that yield conditions of sufficiency for incentives to switch strategy. The main issue concerns individual decision making: When are incentives to switch options sufficient reasons to switch? I advance some principles of sufficiency and draw a general conclusion about incentives and sufficient incentives to switch.[2] Subsequent chapters investigate the type of equilibrium supported by our results.

The decision principle of self-support takes the condition that an option is realized in the same way in which the condition that a profile is realized is

---

[2] The principles advanced are similar to principles proposed in Weirich (1988). The type of argument given for them is new.

taken in the definitions of a solution and an equilibrium, and it involves the same noncompensatory sort of evaluation that those definitions involve (see Section 1.5). Hence it can be used to support the claim that being a strategic equilibrium is necessary for being a solution.

For simplicity, our account of sufficient reasons to switch strategy treats only reasons that are incentives, and only grounds of sufficiency that are related to incentives. An agent may have a sufficient reason to switch that is not an incentive. And an incentive for him to switch may be sufficient on grounds that go beyond incentives. But we put aside all reasons and grounds that are independent of his incentive structure. This chapter's main job is to say when an agent's incentive is a sufficient reason to switch strategy in light of his incentive structure.

4.2. SUFFICIENT INCENTIVES

We are looking for a principle that specifies conditions under which an agent has a sufficient reason to switch from one option to another. We know that an incentive to switch, or a preference for the second option given the first, does not suffice. What must be adduced to obtain a sufficient reason for rejecting the first option? Let us begin by considering the type of principle we want.

Although we want a principle that gives a sufficient condition for a sufficient reason to switch options, the principle need not also give a necessary condition. We may settle for a sufficient condition for a sufficient reason to switch because such a condition yields a necessary condition for an option's being a rational choice, namely, the absence of the sufficient condition. The necessary condition for a rational choice then yields a necessary condition for a solution to a game, that is, being a profile in which each strategy meets the necessary condition for a rational choice given that the profile is realized. The necessary condition of rational choice in this way yields a new equilibrium standard for solutions to games.

There are some pitfalls to avoid. We want a principle that describes a sufficient reason to switch options. Corresponding to such a principle is a principle of rational choice, since the absence of a sufficient reason to switch is a necessary condition for a rational choice. If the necessary condition of rational choice implied by the principle of sufficiency is too strong, so that meeting the necessary condition is not possible in all cases, then the principle of sufficiency fails. If the associated necessary condition of rational choice is too weak, so that it does not eliminate some options as rational choices and does not support intuitions about solutions to "pathological" games such as Matching Pennies in pure strategies, then the principle of sufficiency fails. The associated necessary condition for rational choice should be strong enough to support some intuitions about solutions to "pathological" games. It should not be empty; it should go beyond standard decision principles. Finally, the associated necessary condition for rational choice should be couched in terms of incentives or incentive structures so that it is fairly clear and easy to apply. It should flesh out our account of equilibrium and self-support in a practical way.

This chapter advances a principle of rational choice that fills the bill. The principle describes circumstances under which an incentive to switch is a sufficient reason to switch, or a sufficient incentive for switching, that is, a conditional preference sufficient for rejecting the option not preferred. The argument for the principle reflects the two sides of deliberation: the search for options that are good and the rejection of options that are bad. Necessary conditions for rational choice focus on the rejection of bad options. Since the absence of a sufficient reason to switch is a necessary condition for a rational choice, the presence of a sufficient reason to switch from one option to another is a sufficient reason for rejecting the first option. Although the sufficient reason to switch is conditional on adoption of the first option, the sufficient reason to reject the first option is not conditional, in particular, not conditional on the first option's adoption. A sufficient, conditional reason to switch options is a sufficient, nonconditional reason to reject an option.

The argument for the new decision principle therefore has the following structure: I argue (1) from incentives to switch, or conditional preferences, to the nonconditional rejection of options and (2) from the nonconditional rejection of options to the sufficiency of incentives to switch, or conditional preferences. The argument moves from the conditional to the noncondi-tional, and back to the conditional. The motivation for this argumentative structure is twofold: First, basic reasons are conditional preferences. So we must begin with principles that take us from conditional preferences, or incentives to switch, to nonconditional rejection of options. Conditional preferences are the starting point because only they can be counted on to register the information carried by the assumption that an option is adopted, information that must be considered before an option can be deemed a ra-tional choice. Nonconditional preferences are not a reliable starting point because they do not use all relevant informational resources. They do not use the information carried by the assumption that an option is adopted (unless that option is known to be realized). Second, although a reason to switch obtains conditionally on adoption of an option, the sufficiency of a reason to switch is a nonconditional matter, and so must be supported by circum-stances that obtain nonconditionally. A sufficient reason to switch grounds the nonconditional rejection of an option. A sufficient reason to switch from option A to option B, although it involves the condition that A is adopted, is a reason that unconditionally warrants rejecting A. Although the starting point is conditional comparison, the finishing point is nonconditional rejection – as it must be since the ultimate objective, a decision, is itself nonconditional. So the sufficiency of an incentive to switch, although it concerns a conditional preference, must rest on nonconditional reasons for the rejection of an option.

Let us now begin formulating our principle about sufficient reasons to switch, and the companion necessary condition for rational choice. I first introduce two general principles concerning sufficient reasons for rejecting an option. Then I modify the principles so that the conditions for rejection concern incentives exclusively. The principles state when an incentive to switch from one option to another is sufficient for rejecting the first option. For the first option to be rational, it is then necessary that those conditions not obtain.

A sufficient incentive to switch from option A to option B entails a sufficient reason to reject A. But a sufficient reason to reject A does not entail a sufficient incentive to switch from A to another option since not all reasons are incentives. However, an incentive to switch from A to B is sufficient if and only if it is a sufficient reason to reject A. We use this equivalence to move back and forth from switching options to rejecting options.

An incentive to switch from one option to another makes the first option seem bad and a candidate for rejection. But we have seen that the policy of rejecting each option from which there is an incentive to switch can lead to the rejection of all options, too drastic a result. So our two principles of rejection advance a weaker policy. They work together recursively to reject options. The first principle provides the base case:

If there is an incentive to switch from A to B, and there is no reason to switch from B, then the incentive to switch from A to B is a sufficient reason to reject A.

According to this principle, rejection of A is warranted if there is a preferred alternative that has no opposition. The principle is very plausible since none of the factors that make either conditional or nonconditional preferences insufficient reasons for rejecting options arises in this case. It is not true that for each option there is a more preferred option. And it is not true that given each option another option is preferred.

The second principle provides the recursive step:

If there is an incentive to switch from A to B, and there is a sufficient reason to reject B, then the incentive to switch from A to B is a sufficient reason to reject A.

This principle rejects A because there is a sufficient reason for rejecting an alternative preferred to A given A. It makes sense to reject an option that is worse given the option than an option already rejected. One does not have to worry that the conditional preference is insufficient grounds for rejecting A because of shifting conditions of preference. The first factor in the grounds for rejecting A is a conditional preference, but the second factor is a nonconditional reason for rejecting B. Since the grounds for rejecting A involve only one conditional preference, it cannot be that shifting conditions of preference undermine the grounds. The grounds for rejecting A are secure from shifts of this sort. B's being preferred to A given A is a sufficient reason to reject A in these circumstances since there is nothing to prevent the backward transfer from B to A of sufficiency of reasons for rejection.

These two principles give conditions under which an incentive to switch from A to B is a sufficient reason for rejecting A, or a sufficient reason to switch. But we want principles for sufficiency of incentives that state the grounds of sufficiency in terms of incentives. So next I revise the principles to make them operate on the basis of incentives alone. The revision makes the principles easier to apply. We do not have to consider all reasons, just incentives. Focusing on incentives makes the process of rejection of options relative to reasons that are incentives, however. Using the companion necessary condition of rational choice to formulate an equilibrium standard for solutions therefore makes an equilibrium in a game relative to reasons that are incentives. The identification of an equilibrium with respect to certain reasons is still useful, however, since non-equilibria with respect to those reasons are still disqualified as solutions. Being an equilibrium with respect to those reasons is still a necessary condition for a solution.

The first revised principle again provides the base case:

If there is an incentive to switch from A to B and there is no incentive to switch from B, then the incentive to switch from A to B is a sufficient reason to reject A.

This principle asserts that the absence of incentives to switch from B is enough to make the incentive to switch from A to B sufficient for rejecting A. The absence of all reasons to switch from B is not necessary. The new principle implies that the original principle has a stronger than necessary condition for rejection of A. Suppose that there are some reasons to switch from B that are not incentives. Perhaps there are reasons to switch from B to C because C is part of an equilibrium Pareto superior to an equilibrium containing B. Then B may not be a rational choice. The reasons to switch from B to C may provide reasons for rejecting B. We may conclude that B is a rational choice only if there is no sufficient reason of any type to switch from B. But claiming that the incentive to switch from A to B is sufficient is not the same as claiming that B is a rational choice. It entails only that A is not a rational choice. Although neglecting reasons that are not incentives undermines the case for B, it does not undermine the case against A. If there is no incentive to switch from B, the incentive to switch from A to B is a sufficient reason to reject A. As before, the factors that make incentives insufficient do not arise. It is not true that for each option there is a preferred option. And it is not true that given each option another option is preferred. Given B, no other option is preferred. A's subordination to a rival such as B makes A noncompetitive.

An interesting special case arises if the conditions of the base principle are met and there is also a reason, not an incentive, to switch from B to A. The incentive to switch from A to B, and the reason to switch from B to A, cannot both be sufficient, at least not if A and B are the only options. I claim that the incentive to switch from A to B is sufficient, and consequently the reason to switch from B to A is insufficient. Reasons that are incentives have priority in this context.

The second revised principle again provides the recursive step:

If there is an incentive to switch from A to B, and there is an incentive to switch from B to another option that is a sufficient reason to reject B, then the incentive to switch from A to B is a sufficient reason to reject A.

This principle follows straightforwardly from the original version of the principle. The original version requires a sufficient reason to switch from B, and a sufficient incentive to switch from B certainly qualifies.

Note that when we say that an incentive to switch options is a sufficient reason to reject an option if some conditions are met, we do not mean that if the conditions are met, then the incentive by itself is a sufficient reason to reject the option, but rather that the incentive plus the conditions are sufficient together. So, for example, with regard to the second principle, the sufficient reason to reject A is strictly speaking the incentive to switch from A to B plus the sufficiency of the incentive to switch from B to another option. When we say that the incentive to switch from A to B is sufficient for rejecting A, we mean, strictly speaking, that it is sufficient in the context of the reasons for rejecting B. Logic's principle of detachment does not apply to incentives that are sufficient given some conditions. We cannot detach the sufficiency of incentives from the context with respect to which the incentives are sufficient.

The two revised principles of sufficiency now involve more than one condition of preference, but conditional preferences with differing conditions do not cause trouble for the principles. The basis principle, for instance, entails that if option B is preferred to option A given A, and no option is preferred to option B given B, then A should be rejected. There are two conditions of preferences: realization of A and realization of B. However, the principle holds despite the shift in conditions of preference. The absence of preferences for alternatives to B given B merely grounds the noncondi-tional verdict that A should be rejected on the basis of incentives. The two conditions of preferences are used independently. Likewise in the recursive principle, preferences with respect to B merely ground a nonconditional verdict that A should be rejected on the basis of incentives. Because in both principles the preference ranking with respect to B merely grounds a nonconditional assessment of preferences with respect to A, the shift in conditions of preference does not matter. The two conditions of preference have independent roles.

The new principles of sufficiency require more than an incentive to switch before they yield a sufficient reason for rejecting an option. Since an incentive to switch is all that is usually required, they should seem plausible. The main question about them from our point of view is whether despite the stiffer conditions for rejecting an option, they still reject options too readily, in particular, in the special situations that arise in normal-form games, in which the assumption of an option carries information relevant to preferences between options. That worry can be put to rest. The extra conditions for rejection of an option are strong enough so that even in normal-form games only irrational options are rejected by the principles. Chapter 7 supports this point by examining the consequences of the principles for the normal-form games discussed in Chapter 3.

Notice that in some games, following a sufficient incentive to switch may lower an agent's payoff. Take the ideal three-person normal-form game in Figure 4.1. In addition to Row and Column, a third agent called Matrix chooses between strategies M1 and M2 that yield, respectively, the left and right matrices of rows and columns. The third number in a payoff profile for a row, column, and matrix is Matrix's payoff.

Suppose that M1 is fixed. This yields a subgame for Row and Column with the payoff matrix obtained by removing the third element of each payoff profile in the matrix on the left. The result is in Figure 4.2. (R1, C1) is the unique Nash equilibrium for Row and Column in the subgame obtained by fixing M1. So given M1, we suppose that Matrix believes the profile realized

(1, 1, 1) (0, 1, 1)   (1, 0, 2) (0, 1, 0)

(0, 1, 1) (1, 0, 0)   (0, 1, 0) (1, 1, 0)

Figure 4.1   Rationality penalized.

(1, 1) (0, 1)

(0, 1) (1, 0)

Figure 4.2   The subgame with M1 fixed.

is (R1, C1, M1). That is, he believes that if he does M1, the response of Row and Column is (R1, C1). Given M1 he therefore has an incentive to switch to M2. Now the unique Nash equilibrium in the subgame obtained by fixing M2 is (R2, C2). So for similar reasons, Matrix believes that if he adopts M2, the other agents respond with (R2, C2), and the resulting profile is (R2, C2, M2). This profile is the unique Nash equilibrium of the three-person game. Given Matrix's belief about the response of other agents to M2, he has no incentive to switch from M2 to M1. So Matrix has an incentive to switch from M1 to M2 and no incentive to switch away from M2. This makes the incentive to switch from M1 to M2 sufficient according to our basis principle. But as a result of switching from M1 to M2, Matrix's expectations are worse. This shows that the outcome of following sufficient incentives to switch is not always higher expectations.

The example shows how Newcomb-like phenomena may emerge in normal-form games. If Matrix adopts M1, then if he were to adopt M2, he would be better off. So he has an incentive to switch. It is a sufficient incentive according to our principles. Since Matrix has a sufficient incentive to switch from M1 to M2, it is irrational for him to adopt M1, even though he has lower expectations if he adopts M2 than if he adopts M1. Such conflict between incentives and expectations is the essence of Newcomb's problem. One of the morals of Newcomb's problem is that rationality can be penalized. We take the example of Figure 4.1 as a case in which rationality is penalized. We do not take the example as grounds for rejecting our principles about the sufficiency of incentives.

The two principles of sufficient incentives yield a necessary condition of rational choice. We use some special terminology to present it. Let a *path* of incentives be a sequence of options (possibly infinite) such that for adjacent options there is an incentive to switch from the first to the second. For example, the sequence A, B, C is a path of incentives if there is an incentive to switch from A to B and from B to C. Let us say that a path *terminates* in an option if and only if the option is the last member of the path and there is no incentive to switch from it to another option. Ending is not the same as terminating. A path may end in an option from which there are incentives to switch, but it terminates in an option only if there are no incentives to switch from the option.[3] A path is *closed* if and only if any option in the path that generates an incentive to switch is followed by another option. Every finite closed path terminates in an option. And every terminating path is a closed path that ends.

---

[3] In this section the only type of termination recognized is termination in an option. The next section introduces another type of termination and makes appropriate revisions in the account of terminating paths.

85

We now propose the necessary condition of rational choice:

An option is a rational choice only if there is no terminating path of incentives away from it.

Although a path may have just one option, at which it terminates, such a path does not count as a terminating path *away* from the option. A terminating path away from an option must have at least two members, the initial option and a terminal option. Generally when we speak of a terminating path or say that an option starts a terminating path, we mean a terminating path away from an option, and so a path with at least two members. So generally we say that an option is a rational choice only if it does not start a terminating path of incentives.

The derivation of the preceding necessary condition of rational choice from the two principles of sufficient incentives is straightforward. The basis principle of sufficiency establishes that the second to last option in a terminating path should be rejected. The recursive principle of sufficiency establishes that the third to last option should also be rejected, and so on, for all prior options. Hence every option in a terminating path except the last should be rejected. Consequently no rational option initiates a terminating path of incentives. We do not conclude that the last option of a terminating path, its terminal option, is a rational choice. Several options may serve as terminal options, either of different paths initiated by the same option or of different paths initiated by different options. Some terminal options may be irrational given the availability of the others. The reasons that adjudicate among terminal options are beyond our project's scope, however.

One immediate consequence of our necessary condition for rational choice is that if an option fails to meet it and so starts a terminating path of incentives, then every option in the terminating path, except the last, also fails to meet it. The components of the terminating path undermine its members. To state the point more precisely, let us say that one path is a *subpath* of another if and only if under a renumbering of positions in the first path it is a component of the second path. For instance, the first member of the first path may be the 28th member of the second path. Every closed subpath of a terminating path is also a terminating path (except the terminal option taken by itself). So every closed subpath furnishes a conclusive case against its initial member. The closed subpath shows that its initial member initiates a terminating path.

We identify a path by a series of positions and their occupants. So suppose incentives cycle from A1 to A2 and back endlessly. In the path of incentives A1, A2, A1, ... the first appearance of A1 occupies the first position of the path, and the second appearance of A1 occupies the third position of the path. Each occurrence of A1 precedes a different closed subpath, identified by the positions as well as the occupants of those positions in the whole path.

Cyclical incentives do not produce a cyclical path, but rather an infinite path with repeated appearances of options. The options involved do not form a terminating path.

Let us now show that the proposed necessary condition of rational choice can in fact be satisfied in every decision problem. That is, let us show that in every decision problem there is at least one option that meets the condition. To begin, take an arbitrary decision problem. If any option fails to meet the necessary condition, then there is a terminating path of incentives away from it. The last option of that terminating path meets the necessary condition. Since it is a terminal option, there is no incentive away from it, and so a fortiori no terminating path of incentives away from it. Consequently, not every option of the decision problem can fail to meet the necessary condition. In order for any option to fail to meet the condition, some option must meet it.

4.3. TERMINATION IN A TREE OF INCENTIVES

Suppose that the game of Matching Pennies is embedded in an ideal game with an additional option, F, that is inferior to both Heads and Tails. Think of F as flipping one's penny when such randomization is penalized. Let the payoff matrix be as in Figure 4.3. Suppose that Row believes that Column will respond to F with H, to H with T, and to T with H. These beliefs generate the incentives to switch that Row has given each of his strategies. Given his beliefs, he has the following path of incentives: F, H, T, H, T, ...[4] This path does not terminate in an option and so does not provide grounds for rejecting F according to the preceding necessary condition for rational choice. But Row should reject F.

We want to strengthen the necessary condition for rational choice so that F is ruled out because of the path of incentives it starts. We can do this by revising the principles of sufficiency from which the necessary condition of rational choice is derived, in particular, by generalizing the basis principle for sufficiency of incentives. As formulated previously, the basis principle concerns only incentives to switch from one option to another. But it can be generalized to take account of incentives to switch from an option to a *set* of options. Generalized, it authorizes the rejection of an option because of

---

[4] The incentives in this example are conditional on the strategy adopted, but not on the profile adopted. They are nonrelative incentives, those used to determine nonrelative equilibria. Relative incentives are our chief interest because relative equilibrium is our principal topic. But nonrelative incentives are simpler than relative incentives since relative incentives involve conditionalization with respect to a profile, not just a strategy. The points about sufficient incentives made here for nonrelative incentives carry over to relative incentives. An incentive starting a path of relative incentives with the same structure as a terminating path of nonrelative incentives is a sufficient incentive to switch. Chapter 5 treats paths of relative incentives.

	F	H	T
F	-2, -2	-2, 0	-2, 0
H	0, -2	1, -1	-1, 1
T	0, -2	-1, 1	1, -1

Figure 4.3 Embedded Matching Pennies.

a sufficient incentive to switch from the option to a set of options, even if there is not a sufficient incentive to switch from the option to any member of the set. Then it says that Row has a sufficient incentive to switch from F to {H, T} even if he does not have a sufficient incentive to switch from F to H or from F to T. This section presents the generalization.

Perhaps it seems that the game of Figure 4.3 is better handled by first eliminating strictly dominated strategies and then applying principles already in hand. There are two reasons for not proceeding this way. The first is that the maneuver is ad hoc. We would like a general necessary condition of rational choice that rejects the strictly dominated strategies along with all other strategies for which incentives provide sufficient reasons for rejection. The second reason is that the principle to reject strictly dominated strategies is not well supported. It faces difficulties similar to those facing the standard of incentive-proofness. They are presented at the end of this chapter. For the foregoing reasons, it is better to handle the example by generalizing the basis principle for sufficient incentives.

Our generalization of the basis principle introduces more ways in which a path may terminate, and therefore results in fewer options that do not start terminating paths. The generalization thus strengthens the associated necessary condition for rational choice. The generalization does not excessively strengthen the associated necessary condition, however. As we will show, in every decision problem at least one option meets the strengthened necessary condition for rational choice. Although the generalization of the basis principle raises the hurdle for options, some options can still clear it.

The generalization of the basis principle is inspired by steps in deliberation in which sets of options play a role. The steps we have in mind are moves from consideration of an option A to consideration of a set of options X that does not contain A. In other words, we consider rejection of an option A because of a preference for choosing from a set of options not containing A. The generalization provides conditions under which an incentive to switch from A to X is a sufficient reason for rejecting A. It provides conditions under which an incentive to switch from an option to a set of options is a

88

sufficient reason to switch, that is, a sufficient reason to reject the option.

To begin presenting the generalization, let us consider the psychological nature of an incentive to switch from an option A to a set of options X. We say that *an incentive to switch from A to X exists* if and only if X does not contain A and there is a preference given A for choosing from X instead of choosing A. There may be a preference for choosing from some set X containing A instead of choosing A, but we do not call such a preference an incentive to switch from A to X. Next, to make an incentive to switch from A to X easier to work with, we assume that it obeys a certain principle of rationality, a principle of coherence with incentives to switch options. The principle is met by agents in ideal games, our primary topic. The principle of coherence is, roughly, that an incentive to switch from A to X exists if and only if there is an incentive to switch from A to some option in X. More precisely, taking account of the requirement that X not contain A, it reads as follows:

There is an incentive to switch from an option A to a set of options X if and only if X does not contain A and there is an incentive to switch from A to some member of X.

This principle provides our criterion for the existence of an incentive to switch from A to X.

Our generalized basis principle says that an incentive to switch from A to X is a sufficient reason to switch if X meets certain conditions. To introduce the conditions, I first present what I call a *tree* of paths of incentives, or an incentive tree. A tree is a nonempty set of paths with a common origin. The paths may have shared segments. Our principle is restricted to cases in which there is an incentive to switch from A to a member of X that begins an incentive tree constituted by the members of X. To illustrate, suppose there are incentives to switch from A to B, from B to C, and from B to D. Then there is a tree of paths starting with B and comprising the paths from B to C and from B to D. Figure 4.4 depicts it.

Let X be {B, C, D}. There is an incentive to switch from A to B, a member of X, and X does not contain A. According to our criterion for existence of an incentive to switch from an option to a set, there is an incentive to switch from A to X. A and X meet the restriction since there is an incentive to

Figure 4.4   A tree of incentives.

89

switch from A to B, and B is the first member of a tree constituted by the members of X. Since X comprises the members of a tree T, not containing A, we also say that there is an incentive to switch from A to T. Often we let T double as a name for the set of options in the tree and as a name for the tree itself, a structured set of options. Generally, the distinction between a tree and the set of options it comprises can be ignored.

Suppose that options not including A form a tree T, and there is an incentive to switch from A to the first member of T. By our criterion, there is an incentive to switch from A to T. We propose two conditions under which the incentive is a sufficient reason to switch, or a sufficient reason to reject A. The first condition involves the closure of a tree. I say a tree is *closed* if and only if it is closed under incentives to switch options. More precisely, a tree is closed just in case if there is an incentive to switch from some option in the tree to another option, then wherever the first option appears in the tree, there is a segment in the tree going from the first option to the second. In the example of Figure 4.4 the tree B starts is closed if the incentives to switch depicted are all the incentives to switch. The first condition for the sufficiency of the incentive to switch from A to T is that T be closed. If a tree is closed, a switch to it has the advantage of being final with respect to incentives. Once deliberation switches to the tree, incentives never lead deliberation out of the tree.

The second condition for the sufficiency of the incentive to switch from A to T requires that there be no incentive to switch from a member of T to the first member of a closed tree not containing the member of T. This condition ensures that T is not open to trimming on the same grounds that motivate switching to it. In other words, it ensures that T is a *minimal* tree arising from incentives to switch. If T is closed, there is an incentive to switch from a member of T to the first member of a closed tree T' not containing the member of T if and only if T' is a closed subtree of T, the member of T immediately precedes T', and the member of T does not appear in T'. So in the context of the closure condition, the minimality condition can be stated this way: Every member of T appears in every closed subtree it immediately precedes.

To illustrate how the minimality condition may be violated given satisfaction of the closure condition, suppose that there is an incentive to switch from A to B, from B to C, from C to D, and there are no other incentives, in particular, no incentives to switch from D to another option. Then the path B, C, D is a closed tree, and there is an incentive to switch from A to it since A does not appear in it. But the tree does not satisfy the minimality condition. There are a member B and a closed subtree C, D such that the member immediately precedes the subtree and does not appear in it.

The minimality condition has intuitive appeal and also accommodates infinite trees. Consider the case of picking a natural number that gives one's income. Each natural number n precedes a closed tree starting with $n + 1$ and excluding n. There is an incentive to switch from n to the closed tree starting with $n + 1$. If sufficiency did not require minimality in addition to closure, there would be a sufficient incentive to switch from n to the closed tree starting with $n + 1$. Consequently, there would be a sufficient incentive to switch from every natural number to the set of its successors. Hence every natural number would be rejected. There would be no rational choice in the decision problem, contrary to intuition. The minimality condition prevents this counterintuitive result. The incentive to switch from n to the closed tree beginning with $n + 1$ does not meet the minimality condition. For there is also an incentive to switch from $n + 1$ to the closed subtree beginning with $n + 2$. Hence $n + 1$ fails to appear in a closed subtree it immediately precedes.

In view of the foregoing considerations, we adopt the following generalization of the basis principle:

If there is an incentive to switch from an option A to the first member of a closed tree T not containing A, and every member of T appears in every closed subtree it immediately precedes, then the incentive to switch from A to T is a sufficient reason to reject A.

If there is an incentive to switch from A to T, then T does not contain A but contains an option preferred to A given A. If T is also closed, then deliberation has no incentive to return to A once it switches attention to T. The dismissal of A is final. If the minimality condition is also met, then there is no incentive to switch from a member of T to a closed subtree T. T is reduced as far as possible with respect to incentives to switch from options to closed trees. Thus T is the final result of bringing such incentives to bear on deliberations that start with A. The twofold finality of the incentive to switch from A to T makes it a sufficient reason to reject A.

Notice that the new version of the basis principle for sufficiency of incentives entails the old version. According to the old version, an incentive to switch from A to B is sufficient if there is no incentive to switch away from B. Suppose then that there is an incentive to switch from A to B and no incentive to switch away from B. According to the old version of the basis principle, the incentive to switch from A to B is sufficient. Now let T be the tree comprising B and B alone. Given the incentive to switch from A to B, there is an incentive to switch from A to the first member of T. T does not contain A since B must be distinct from A in order for there to be an incentive to switch from A to B. There is, then, an incentive to switch from A to T. Also, T is closed under incentives to switch since there is no incentive to switch away from B. Furthermore, no member of T immediately precedes a closed subtree in which the member does not appear – T has no proper subtrees. So the incentive to switch from A to T is sufficient according to the new version of the basis principle. Since an incentive to switch from A to T is equivalent to an incentive to switch from A to B, the incentive to switch from A to B is sufficient according to the new version of the basis principle, just as it is according to the old version. Thus if an incentive to switch from one option to another is sufficient according to the old version of the basis principle, it is sufficient according to the new version as well.

The new version of the basis principle yields a new, stronger necessary condition for rational choice when taken together with the recursive principle. To present the necessary condition, we define a path of incentives as before – that is, as a sequence of options (possibly infinite) such that for adjacent options there is an incentive to switch from the first to the second – except that now we allow a path to have a tree as its last member. That is, a path, identified by a series of positions and occupants of the positions, may have a tree as the occupant of the last position. If the terminal tree is itself a path of options, then for our purposes it is equivalent to, although not identical to, a path in which every position is occupied by an option.

When a tree is the last member of a path, there must be an incentive to switch from the penultimate member of the path to the first member of the tree. Suppose that the tree does not contain the penultimate member of the path so that there is an incentive to switch from the penultimate member to the tree itself. We say then that the path *terminates in the tree* if and only if the incentive to switch from its penultimate member to the tree meets the conditions for sufficiency given, namely, the conditions of closure and minimality. This is the only sort of termination in a tree we recognize. A path terminates in a tree if and only if there is an incentive to switch from the penultimate member of the path to the tree and the incentive is sufficient according to the new version of the basis principle.

Termination in a tree subsumes termination in an option. Since an option is itself a degenerate tree comprising a single option, every multimembered path ends in a tree such that there is an incentive to switch from the penultimate member of the path to the first member of the tree. Suppose the path terminates in a option. Then the old basis principle is responsible for the rejection of the penultimate option. Since, as we showed previously, the new basis principle entails the old, the new principle is also responsible for the rejection of the penultimate option. But in that case the path terminates in the tree comprising the terminal option.

We distinguish between a path's *ending* in a tree and its *terminating* in a tree. Consider three distinct options A1, A2, and A3, and the sequence of incentives A1, A2, A3. We may take this sequence as the path that has A1 as its first member and the tree A2, A3 as its second member, since there is an incentive to switch from A1 to the first member of that tree. Then the path *ends* in the tree A2, A3. But the path does not *terminate* in the tree if the tree is not closed – if, for instance, there is an incentive to switch from A3 to A4. In that case the incentive to switch from A1 to the tree A2, A3 does not meet the closure condition for sufficiency that appears in the definition of termination.

We say that a path is infinite, even if it has only a finite number of members, if the last member is an infinite tree. Termination in a tree makes it possible for an infinite path to terminate. As before, a closed finite path terminates, but now not every closed path that terminates is finite. Some closed paths terminate in infinite trees. Moreover, an infinite path may end in an infinite tree, so ending is no longer a sign of a finite path. Ending in an infinite tree, of course, differs from termination in an infinite tree. Consider a path that ends in an infinite tree that repeats every option in the path. In this case there is no incentive to switch from the penultimate member to the tree itself, since the tree contains the penultimate member. So the path does not terminate in the infinite tree.

The foregoing tells us how to determine whether a sequence of incentives terminates once the incentives have been organized into a path. But now that a tree is allowed as a final member of a path, there are several ways of organizing a sequence of incentives into a path. Consider the sequence of incentives A1, A2, A3, ... It may be construed as a path of n positions with A1 occupying the first position, A2 occupying the second position, and so on, or as a path of two positions with A1 occupying the first position and the tree A2, A3, . . . occupying the second position. There are other possibilities as well. We say that a terminating path away from an option exists if there is any way of organizing incentives to form a terminating path away from the option. We do not assume a particular way of organizing sequences of incentives into paths.

The revised principles of sufficient incentives yield a new necessary condition for rational choice, which has the same formulation as before:

An option is a rational choice only if there is no terminating path of incentives away from it.

However, the content of the condition is different now that termination in a tree is allowed.

The revised principles of sufficient incentives enable us to derive the new necessary condition as before. The basis principle rejects the option appearing second to last in a terminating path away from an option, and the recursive principle rejects prior options one by one until the first option in the path has been rejected. If a path terminates in either an option or a tree, incentives furnish sufficient reasons to reject all the options before the terminus. Therefore, a rational choice does not start a terminating path.

To see that the addition of termination in a tree yields a stronger necessary condition for rational choice, let us return to the game of Figure 4.3, Embedded Matching Pennies, which began this section. Row believes that Column will respond to F with H, to H with T, and to T with H. So Row has the following path of incentives: F, H, T, H, T, ... The path does not terminate in an option and so does not provide grounds for rejecting F according to the first version of the necessary condition for rational choice. But the path may be taken to end in the tree comprising the H–T cycle since there is an incentive to switch from F to the first member of the tree. Also, there is an incentive to switch from F to the tree, since the cycle does not contain F. Moreover, the cycle is closed under incentives to switch, and it meets the minimality condition since for every member, every closed subtree the member immediately precedes contains the member. Thus the path terminates in the tree comprising the cycle. Hence, according to the new, strengthened version of the necessary condition for rational choice, the terminating path provides grounds for rejecting F. Although F does not initiate a path terminating in an option, it does initiate a path terminating in a tree.

Although the new necessary condition is stronger than the old, it is not excessively strong. It does not reject any option that is in fact a rational choice. Intuition confirms that the inferior options rejected in the game of Figure 4.3 are not rational choices. The minimality condition, moreover, prevents the rejection of rational choices in the problem of choosing your income. In general, if pursuit of incentives leads away from an option A to a closed tree T not containing A, and pursuit of incentives does not trim T further, then A is not a rational choice. The only remaining doubts about the necessary condition concern the incentives with respect to which it ought to be applied. We address this issue in the next section.

Before closing this section, let us verify that the strengthened necessary condition can be met in all cases. We want to show that in every decision problem there is at least one option that meets the new, stronger necessary condition. That is, we want to show that in every decision problem there is an option that does not initiate a path terminating in a tree. Showing this also shows that the option does not initiate a path terminating in an option since, as we saw earlier, termination in an option is subsumed by termination in a tree.

The proof is similar to the proof for termination in an option. Take a decision problem. Suppose by way of *reductio* that each option initiates a path terminating in a tree. Select an option, and let T stand for the tree that terminates the path the option starts. Consider the members of T. By hypothesis, each initiates a path terminating in a tree. Select one of these paths and call its terminus T'. Since T and T' are terminal trees, and terminal trees are closed under incentives to switch, T' is a closed proper subtree of T, and the option that immediately precedes T' is a member of T. Also, since T' is

94

a terminal tree, there is an incentive to switch to it from the option immediately preceding it, and so T' does not contain that option. But then T' and its immediate pred ecessor constitute a violation of the minimality condition for T. This contradicts the assumption that T is a terminal tree and shows that our initial hypothesis is false. Some option fails to initiate a terminating path.

4.4. TERMINATION WITH RESPECT TO PURSUED INCENTIVES

The case against an option that starts a terminating path of incentives appeals to the results of pursuing those incentives, namely, switching options and never returning to the original option. But not all incentives are pursued. If an agent has several incentives away from an option, he pursues at most one. And if pursuit of an incentive is futile, a rational agent may not pursue it. In particular, a rational agent may not pursue an incentive to switch if the closed path it initiates does not terminate. He may ignore a path that goes nowhere. The case against an option that starts a terminating path of incentives needs revision to take account of nonpursuit of incentives. If an option starts a terminating path of incentives, but some incentives in the path are not pursued, the path does not constitute a sufficient reason to reject the option. Our necessary condition of rationality for an option, no terminating path away, has been formulated in terms of an agent's incentives to switch whether he pursues them or not. It should be revised to take account of the selectiveness of an agent's pursuit of incentives.

Strictly speaking, the revision should take account of an agent's *information* about his pursuit of incentives. But the necessary condition is for an ideal agent, who has full knowledge of his pursuit of incentives. Such an agent knows whether he pursues any incentive to switch away from an option and knows which of his incentives to switch away from an option he pursues if he pursues any. For an ideal agent, revising to take account of his information about his pursuit of incentives is the same as revising to take account of his pursuit of incentives. So the distinction between pursuit of incentives and information about it can be ignored.

4.4.1. Pursuit of Incentives

To set the stage for a revision that takes account of nonpursuit of incentives, let us explain more carefully pursuit of an incentive to switch options. Pursuit of an incentive to switch options, in our technical sense, is not actual choice. It is a nontemporal analogue for single-stage games of real, temporal pursuit of an incentive to switch options in multistage games. In our technical sense, pursuit of an incentive to switch indicates a hypothetical choice, that is,

95

a disposition to make a certain choice given some assumptions about the choice. We say that *an agent pursues an incentive to switch from option A to option B* if and only if B is realized if either A or some option to which the agent has an incentive to switch from A is realized, in other words, if and only if B is chosen given either A or some option preferred to A given A.

To pursue an incentive to switch is to reject an option in favor of another that is conditionally preferred when choice is restricted to the set comprising the option and all other options conditionally preferred. Pursuit of the incentive indicates a hypothetical choice among the option and the options to which there is an incentive to switch given the option. The restricted set of options for the hypothetical choice includes all rivals to the option, however many there are. But the restricted set is generally smaller than the set of all options.

A hypothetical choice behind pursuit of an incentive does not involve a new, hypothetical decision problem. It pertains to the actual decision problem. The hypothesis is that the choice in the actual decision problem falls within a certain subset of options. It is a hypothesis about the outcome of the choice. It is not a restriction on the options available. Claiming that an incentive is pursued is claiming that if the choice is in a certain subset of options, it is a certain option of the subset.[5]

The assumption that the choice is in a certain subset of options does not involve an assumption that the agent's information changes. There is no change in the availability of options, for instance, to justify a change in information. The agent's information is the same as in his actual decision problem. If the choices within the subset of options are irrational given the agent's information, then even if he is ideal, his information does not change to make one of the choices rational. His hypothetical choice just indicates which of those irrational choices he makes given that he makes one. In such a case an ideal agent's hypothetical choice exhibits a type of damage control. It involves a minimal departure from rationality. It is a rational way of choosing given a choice that falls in a set of irrational options. When we say that an ideal agent makes his hypothetical choice rationally, we mean only that he is as rational as the hypothesis allows.

Hypothetical choice is taken to restrict the outcome reached rather than the options available because it serves as the basis for application of the principle of self-support presented later. If it restricted the options available, and thereby introduced a new decision problem, application of the principle of self-support, which makes actual choice depend on hypothetical choice, would start a regress. According to the principle, rational choice in a decision

---

[5] Because a hypothetical choice of this type involves no change in the decision problem, it involves no change in the grounds of preference among options. Hence the problems with principle alpha discussed in Section 4.5 do not arise.

problem would depend on rational choice in a new decision problem with a restricted set of options. Likewise rational choice in the new decision problem would depend on rational choice in another decision problem with an even more restricted set of options, and so on. In cases in which the original set of options is infinite, the regress may be infinite despite the reduction of the set of options at each step of the regress. Our interpretation of hypothetical choice allows us to cut off such regresses.

Hypothetical choice with respect to an option and its rivals provides a means of formulating a necessary condition of rational choice. Although pursuit of an incentive from A to B is not the same as rejection of A, hypothetical choices with respect to subsets of options impose constraints on actual choice with respect to the set of all options. For a rational ideal agent, pursuit of an incentive from A to B amounts to rejection of A. We revise the necessary condition of rational choice presented in the previous section to express this connection between pursuit of incentives and rational choice. Instead of saying that a rational option does not initiate a terminating path of incentives, we say that it does not initiate a terminating path of pursued incentives. That is, we say the following:

An option is rational only if it does not start a terminating path of incentives actually pursued.

To clarify the revision, let us consider an example, which presents an agent's pursuit of incentives in two stages: The first stage presents the incentives pursued given relentless pursuit of incentives, that is, given pursuit of incentives at every opportunity. For each option generating incentives to switch, this stage designates the incentive to switch pursued if any is.[6] The second stage presents the incentives pursued, putting aside the assumption of relentless pursuit of incentives.

Take the ideal normal-form game in pure strategies in Figure 4.5. To describe pursuit of incentives in this game, let us introduce some terminology. Let us call an incentive to switch from an option A to an option B *optimal* if and only if given A, among options to which there is an incentive to switch, B has maximum expected utility. Suppose that Column pursues incentives at every opportunity; Row and Column both pursue optimal incentives if they pursue any; and given C2 Row breaks the tie between R1 and R2 in favor of R2. Each agent knows the way the other agent pursues incentives and uses this information to calculate his own incentives. Take Row's perspective.

---

[6] Pursuit of an incentive away from an option given relentless pursuit of incentives is a hypothetical choice with respect to a restricted set of options comprising the option and options conditionally preferred to it, given that a conditionally preferred option is chosen. The hypothetical choice assumes first that the outcome is restricted to an option and its rivals and then that the outcome is restricted to its rivals. The choice is doubly hypothetical.

$$1, 2 \qquad 1, 1$$

$$2, 2 \qquad 1, 1$$

$$3, 1 \qquad 0, 2$$

Figure 4.5   Pursuit of incentives.

Figure 4.6   The start of a closed tree.

He has no path away from R1 that terminates in a strategy. If he adopts R1, he predicts C1. So he has a path from R1 to R2. He also has a path from R1 to R3. The closures of these paths – the paths formed by including incentives to switch wherever they exist – continue endlessly, repeating strategies. Hence none of them terminates in a strategy.

Row also has no path away from R1 that terminates in a tree. Row has closed paths away from R1 that do not repeat R1, but cycle from R2 to R3 and back to R2. However, he also has closed paths away from R1 that repeat R1. They go to R3 and then to R1. Given R3 Row foresees that Column does C2, and given C2 Row has an incentive to switch to R1 as well as an incentive to switch to R2. Because of the paths away from R1 that return to R1, none of the paths away from R1 terminates in a tree. In order for a path away from R1 to terminate in a tree, it must eventually reach R2 or R3 and then terminate in the closed tree initiated by that strategy. The closed tree beginning with R3 starts as in Figure 4.6. Since that tree contains R1, a path from R1 that reaches R3 does not terminate in the tree; the tree eventually includes its immediate predecessor, since the path from R1 to the tree's immediate predecessor recurs in the tree after R1. Likewise, the closed tree beginning with R2 contains R1. It has a branch that goes to R3 and then to R1. So a path from R1 that reaches R2 does not terminate in that tree; the tree also eventually includes its immediate predecessor.

Since Row has no terminating path of incentives away from R1, R1 meets Section 4.3's necessary condition for rational choice. This happens because incentives not pursued are included in paths of incentives. Row pursues the incentive to switch from R3 to R2, but not the incentive to switch from R3 to R1. That is, among his incentives to switch from R3, he selects the incentive to switch from R3 to R2. Nonetheless, the incentive to switch from R3 to R1 contributes to a path of incentives from R1 back to R1. The existence of the path back to R1 precludes termination of any path away from R1. No path away from R1 terminates in a profile, and none terminates in a tree because of the path from R1 back to R1. Every closed path R1 starts is endless and includes R3, from which it is possible to get back to R1.

Despite the absence of a terminating path away from R1, there is a good case for rejecting R1 as a rational choice. Since Row knows how he pursues incentives, he knows that if he switches from R1 to R3 he does not switch back to R1. He knows that switching from R1 to R3 is not futile. Granting that the futility of a switch is the only reason for the insufficiency of an incentive to make the switch, Row's incentive to switch from R1 to R3 is sufficient. Row should reject R1.[7]

To make the case more vividly, let us consider Row's pursuit of incentives. If Row pursues incentives relentlessly, then his pursuit of incentives away from R1 goes as follows: R1, R3, R2, R3, R2, ... In other words he pursues incentives endlessly. But endless pursuit of incentives is impossible. Row does not actually pursue incentives relentlessly; Row stops pursuit of incentives in the path at some point. It cannot be true for each pair of adjacent options that if an option in the pair is realized, it is the second. For some pair, despite the incentive to switch from the first to the second, the first is realized.[8] Let us suppose that Row settles on a strategy in the cycle involving R3 and R2. Let us say it is R2. Row knows that Column adopts C1 in response to R2,

---

[7] Recall that speaking of incentives to switch is just a figurative way of speaking of conditional preferences. Strictly speaking, the claim that switching from R1 to R3 is not futile is a claim that conditional preferences, and the hypothetical choices to which they lead, have a certain structure. And the claim that the incentive to switch from R1 to R3 is a sufficient reason to switch is a claim that the conditional preferences and hypothetical choices have a structure that warrants rejecting R1.

[8] Since pursuit of incentives is defined in terms of hypothetical choices, and since hypothetical choices may violate rationality constraints, it is possible that an irrational agent pursues incentives endlessly. In this case relentless pursuit of incentives conflicts with choice. The realization of an option contravenes the agent's pursuit of incentives. That is, his actual choice is incoherent given one of his hypothetical choices. However, ideal agents such as Row do not have hypothetical choices that collectively are in conflict with making an actual choice. Hypothetical choices that collectively conflict with actual choice are irrational. So in ideal games relentless pursuit of incentives, although psychologically possible, is not actual. Agents comply with principles of rationality connecting actual choices with hypothetical choices with respect to subsets of options. Their actual choices cohere with their hypothetical choices.

Since ideal agents do not pursue incentives endlessly, one may wonder how the assumption that pursuit of incentives is relentless should be taken for ideal agents. Suppose that the assumption requires them to pursue incentives endlessly. What happens? When we consider relentless pursuit of incentives in ideal games, we make the idealization of rationality subordinate to the assumption of relentless pursuit. In other words, we let the agents violate rational constraints connecting hypothetical and actual choices. We give priority to the assumption of relentless pursuit of incentives over the assumption of rationality in which it is embedded.

and so he has an incentive to switch from R2 to R3. But he fails to pursue the incentive to switch, seeing that the switch is futile. He knows that if he adopts R3, then Column adopts C2, and he has an incentive to switch back to R2.

Row foresees that pursuit of his incentive to switch from R1 to R3 leads to his adopting R2, since he knows the other incentives he pursues. So he knows that pursuit of the incentive to switch from R1 to R3 is not futile. Although the incentive initiates nonterminating paths of incentives, not all of the incentives in the paths are pursued. Pursuit of the incentives in the paths stops with R2.[9] We take Row's incentive to switch from R1 to R3 as a sufficient incentive to switch. An agent's incentives to switch depend on the incentives he pursues since these trigger responses of other agents. His pursuit of incentives affects the sufficiency of his incentives. In an ideal game if an incentive is not pursued, we claim that it can be ignored in calculating incentives. The sufficiency of an agent's incentives should be assessed with respect to the agent's selection of incentives to be pursued. The justification is that rationality requires agents to acknowledge only incentives that are pursued in computing their incentives. This justification is noncontroversial when it comes to an agent's ignoring incentives of other agents that he knows they do not pursue. Our view is that it is appropriate for an agent likewise to ignore incentives of his that he knows he does not pursue. He is part of the situation in which he decides. Facts about himself set the context for his decision. If he knows that he does not pursue a certain incentive, he should use that knowledge in calculating incentives for other agents and then for himself. Since other agents take account of his hypothetical responses to them, he should take account of his hypothetical responses to them. Rational strategic reasoning requires him to consider his hypothetical responses to them; it is part of considering their hypothetical responses to him. Taking Row's incentive to switch from R1 to R3 as sufficient just puts Row's information about himself on a par with his information about Column. It has him treat his pursuit of incentives after his switch the same way he treats Column's pursuit of incentives after his switch.[10]

Generalizing on the example, we obtain an argument for taking an incentive to be sufficient if it initiates a terminating path with respect to the incentives the agent pursues. The argument goes as follows: An incentive is sufficient if its pursuit is not futile, and its pursuit is not futile if there is no return to the option, given actual pursuit of incentives, even if unpursued incentives return to the option. In other words, the incentive's pursuit is not

---

[9] Strictly speaking, pursuit of incentives stops at a certain position in the paths occupied by R2, but the exact position where stoppage occurs has no practical significance, so we ignore it.

[10] Sobel (1976, 1982) adopts a similar point of view when applying utilitarian principles to cases in which an agent has made or will make a mistake.

futile if given actual pursuit of incentives it leads to a terminal option or set of options. The futility of pursuit of incentives in games depends on the response of an agent to other agents as well as the response of other agents to him. As a result, the sufficiency of an incentive depends on what happens if it is pursued, in particular, what happens in virtue of the agent's nonpursuit of other incentives.

To express the point another way, take an ideal agent's *incentive structure*, that is, the collection of paths emanating from all his options, and trim away the incentives not pursued. If in the trimmed incentive structure an option starts a terminating path, pursuit of the incentives of that path is not futile. There is no return to the option via those incentives. The initial switch away from the option is final. The terminal option or set of options is an alternative more choiceworthy than the option. A terminating path in the trimmed incentive structure therefore indicates that the initial option is not a rational choice.

Our necessary condition for rational choice, revised according to the foregoing view on sufficient incentives, states that an option is rational only if it does not start a terminating path of incentives actually pursued. The necessary condition for rational choice takes the sufficiency of an incentive to be dependent on the incentives an agent pursues. Changes in the agent's pursuit of incentives may transform insufficient incentives into sufficient incentives, or vice versa. If an agent pursues incentives in an irrational way or irrationally stops pursuit of incentives, the influence of his pursuit of incentives on the sufficiency of his incentives is more complex than the necessary condition assumes. The rationality of options has to compensate in complex ways for the irrationality of his pursuit of incentives. So our statement of the necessary condition assumes that an agent's pursuit of incentives is rational.

The standard way of justifying this assumption is to advance it as an idealization. Our necessary condition is a principle of rational choice, or a decision principle. Decision principles generally assume idealizations. They govern rational choice for an agent who is ideal. For nonideal agents, excuses for failing to meet the principles are possible. Theorists put aside these excuses by assuming an ideal agent. The difficult job of handling the irrationalities of nonideal agents is left to auxiliary accounts of rationality. We want to advance our necessary condition's assumption of rational pursuit of incentives as an idealization. However, some questions arise: Is the assumption of rational pursuit of incentives an appropriate idealization for a decision principle? Doesn't rational pursuit of incentives amount to rational choice? And isn't that an inappropriate idealization for a decision principle?

The assumption of rational pursuit of incentives is satisfied by an ideal agent of game theory, who is fully rational as well as cognitively unlimited, and so meets all standards of rationality. Even the assumption of a rational choice is an appropriate idealization for an agent in an ideal game since other agents depend on his rationality. But the ideal agents of decision theory are ideal in a slightly more limited sense than the ideal agents of game theory. In decision theory, an ideal agent is in a fully rational state of mind and is cognitively unlimited. In consequence, the beliefs and desires on which his choice rests are rational, and he has full knowledge of all logical and mathematical truths. But he is capable of making an irrational choice. Otherwise decision principles presume what they set out to explicate. Our necessary condition of rational choice should not presume that the agent chooses rationally, as in game theory. It should be formulated specifically for ideal agents in the decision-theoretic sense. Must it therefore be revised further?

One way to avoid further revision is to exploit the distinction between rational choice and rational pursuit of incentives, or rational choices in hypothetical situations. Since rational pursuit of incentives is distinct from rational choice, we can assume rational pursuit of incentives as input for our decision principle without assuming rational choice. The idealization for agents in decision theory can include the rationality of pursuit of incentives without also including the rationality of the choice itself. We take this approach. Our decision principle is in fact suitable for agents ideal in a decision-theoretic sense. Their pursuit of incentives is rational even if they are capable of making an irrational choice. Their ideality entails rational pursuit of incentives but not rational choice.[11]

Our revised necessary condition of rational choice appeals to termination of paths of incentives actually pursued. Termination of a path of incentives is defined as before, but now it is calculated with respect to a reduced structure of incentives. The relevant structure of incentives is obtained by first removing incentives that the agent does not pursue even if he rationally pursues incentives at every opportunity, and then removing incentives that the agent does not pursue if he rationally abandons pursuit of some incentives. I call the structure resulting from the first step the agent's *partially* reduced incentive structure, and the structure resulting from the second step the agent's *completely* reduced incentive structure. In the partially reduced incentive structure, trees of incentives are reduced to paths, termination in a tree is termination in a subpath, and a terminal subpath with a finite number of options is a cycle. Termination in a subpath in the partially reduced incentive structure becomes termination in an option in the completely reduced incentive structure. In the completely reduced incentive structure, all paths

---

[11] When we say that an ideal agent pursues incentives rationally, we mean he complies with the general standards for rational pursuit of incentives. This does not entail that he complies with the general standards for rational choice. There are, however, two-option decision problems that conflate the distinction between hypothetical choice among a subset of options and actual choice among all options. In these decision problems rational pursuit of incentives entails rational choice *given* the problems' special features.

terminate, so an option does not start a terminating path if and only if it does not start a path. Therefore, the new necessary condition goes as follows:

An agent's option is a rational choice only if there is no incentive to switch away from it in the agent's completely reduced incentive structure.

For clarity, when we say that some incentive is sufficient, we often specify the incentive structure to which its sufficiency is relative: the unreduced, the partially reduced, or the completely reduced incentive structure. The unreduced and partially reduced incentive structures serve only to explain the completely reduced incentive structure. What matters for our necessary condition of rational choice is sufficiency with respect to the completely reduced incentive structure. Incentives sufficient with respect to it are sufficient, period. Some incentives do not initiate a terminating path in the unreduced or partially reduced incentive structures, but do in the completely reduced incentive structure. We intend sufficiency relative to the completely reduced structure when we say that a necessary condition of a rational choice is the absence of sufficient incentives to switch. An option is rational only if it does not generate a sufficient incentive to switch, that is, only if it does not initiate a terminating path in the completely reduced incentive structure.

The completely reduced incentive structure is obtained from the unreduced incentive structure by selecting an incentive to pursue where there are multiple incentives and by selecting a place to stop pursuit of incentives where relentless pursuit of incentives is endless. It does not discard all incentives ignored by the agent's choice. In the case of an ideal, rational agent of game theory, using information about the agent's choice introduces the assumption that his choice is rational. So an incentive structure taking account of his choice cannot be used to justify his choice as rational. The completely reduced incentive structure is used to justify the agent's choice. It does not use information about the agent's choice.

In the completely reduced incentive structure, incentives not pursued are removed. But not all incentives are removed. An agent's completely reduced incentive structure has paths of incentives. An incentive away from the option chosen is not removed from the completely reduced incentive structure simply because of the choice. An irrational agent may adopt an option that starts a terminal path in his completely reduced incentive structure. Although in our technical sense he pursues an incentive away from the option, he nonetheless adopts the option. Although no incentive away from the option adopted is acted on, an incentive away from the option adopted can be pursued in our technical sense because such pursuit concerns hypothetical choices with respect to reduced sets of options, whereas adoption of an option concerns actual choice with respect to all options. In our technical sense pursuit of an incentive away from an option merely indicates that another option is chosen given hypothetically that the choice falls in the set comprising the original option and the options conditionally preferred to it. The completely reduced incentive structure includes pursued incentives ignored by actual choice. It indicates dispositions to choose when choice falls within subsets of options, and not dispositions to choose when choice falls within the whole set of options.

Our necessary condition of rational choice yields our principle of self-support. The principle to choose an option that does not initiate a terminating path says to avoid a certain type of self-defeat. An option that is not self-defeating we call self-supporting. So the principle to meet the necessary condition says to attain a certain type of self-support; we call it the *principle of self-support*:

Choose an option that does not start a terminating path of incentives in your completely reduced incentive structure.

An option's being self-supporting, and so rational, requires that the option not start a terminating path of incentives in the agent's completely reduced incentive structure. Self-support requires the absence of sufficient incentives to switch in that structure.

If reasons are restricted to incentives and failure to initiate a terminating path makes an incentive insufficient, a strategy is self-supporting if it does not initiate a terminating path of incentives. But if reasons are unrestricted, a strategy that does not initiate a terminating path only meets a necessary condition for being self-supporting: There is no sufficient reason of a certain type for switching away from it. We say that a strategy is *strategically* self-supporting if and only if on grounds restricted to incentives there is no sufficient incentive to switch from it to another strategy given that the first strategy is realized. Strategic self-support is the foundation of our new type of equilibrium, strategic equilibrium. Since we are interested mainly in strategic self-support, we often say for brevity that a strategy is self-supporting without specifying that its self-support is strategic.

We have advanced only one set of conditions for sufficiency of incentives. The previous section made them more comprehensive by accommodating a path's termination in a tree. Still the conditions for sufficiency of incentives may not be exhaustive. To put aside this issue, we restrict ourselves to games where no other conditions make incentives sufficient. We assume that in the games we treat unless an incentive starts a terminating path, it is insufficient. The restriction allows us to take starting a terminating path as a characterization or working definition of sufficiency for incentives.

The principle of self-support says that an option is a rational choice only if there is no sufficient incentive to switch away from it. Sufficiency is assessed with respect to the agent's completely reduced incentive structure.

We do not say in general that failure to pursue a sufficient incentive is irrational since there may be multiple sufficient incentives away from an option, and pursuit of at most one is possible. However, where there are sufficient incentives to switch, rationality requires pursuit of one of them. Moveover, if an incentive is sufficient with respect to the completely reduced incentive structure, where unpursued incentives have been discarded, then rationality requires its pursuit.

Some may worry that the new necessary condition and the principle of self-support that follows from it are too demanding. The incentives of the completely reduced incentive structure are all sufficient. The paths they initiate all terminate. Since self-support is defined with respect to an agent's completely reduced incentive structure, and since an option is self-supporting just in case it does not start a terminating path, self-support requires the absence of incentives to switch in the completely reduced incentive structure. The principle of self-support is thus the standard of incentive-proofness for individual choice applied to the completely reduced incentive structure. It is not incentive-proofness, but it is pursued-incentive-proofness. In some games, such as an ideal version of Matching Pennies, at least one agent is unable to meet the standard of incentive-proofness. For some agent, the option realized results in an incentive to switch. Does a similar problem of unattainability arise for the standard of pursued-incentive-proofness? Is a weakening of the standard necessary?

Despite similarity to the standard of incentive-proofness, which we rejected as too strong, our necessary condition of rational choice can be met in every decision situation. Meeting the condition requires an option that does not generate incentives to switch in the completely reduced incentive structure, not the unreduced incentive structure. This makes all the difference. In every decision situation an agent has some option from which there are no incentives to switch in his completely reduced incentive structure. This result can be shown in two easy steps: First, we use Section 4.3's proof that every *unreduced* incentive structure has an option that does not initiate a terminating path to show that every *completely reduced* incentive structure has an option that does not initiate a terminating path. The earlier proof suffices since it does not impose any restrictions on the form of an unreduced incentive structure. Next, we point out that in a completely reduced incentive structure, every path terminates. Hence some option does not start any paths at all. It generates no incentives to switch in the completely reduced structure. It is incentive-proof with respect to pursued incentives.

To obtain the same result more directly, we can reason as follows: If a path terminates in an option, its terminal option does not start a path that terminates in an option. So not every option starts a path that terminates in an option. In the completely reduced structure, every path terminates in an option. So not every option starts a path. Therefore, every completely reduced incentive structure has an option that does not generate incentives to switch, and so is incentive-proof with respect to pursued incentives. We conclude that the new necessary condition and the principle of self-support can be satisfied in every decision problem, and so are not too strong.

4.4.2. *Rational Pursuit of Incentives*

Some may object that the principle of self-support is circular: It states that rationality requires pursuit of sufficient incentives, but an incentive's sufficiency depends on the direction of rational pursuit of incentives. The short response to this objection is that the principle says only that rational choice depends on rational pursuit of incentives. This is not circular. But a more adequate response requires showing that the generation of a rational pattern of pursuit of incentives is not circular. The rest of this section is devoted to a partial account of rational pursuit of incentives that removes the specter of circularity. It presents some principles of rationality for pursuit of incentives. The principles elucidate rational pursuit of incentives and help fill out examples in which pursuit of incentives is claimed to be rational so that claims made about self-support and equilibria in the examples are better supported.

Our approach to rational pursuit of incentives claims that nonpursuit of an incentive is rational only if the incentive is insufficient. Since sufficiency and pursuit are interdependent, we have to be careful about circularity. To prevent circularity, the account of sufficiency of incentives has to appeal to some incentive structure besides the completely reduced structure. Our main suggestion is to look at where an incentive leads if it is pursued in order to show that it need not be pursued. The simplest proposal along these lines is to restore the incentive to the completely reduced structure and see whether it starts a terminating path in the augmented structure. However, this procedure for assessing sufficiency does not adequately handle the dependency of sufficiency on pursuit of incentives. It treats sufficiency as a local rather than a global feature of rational pursuit of incentives. It assesses incentives one by one. We need a method of explaining the rationality of pursuit of incentives that acknowledges that whether an incentive is sufficient depends not just on what happens if that incentive is pursued, but also on whether other incentives are pursued. To capture the rich, global structure of justification of pursuit of incentives, we take a two-stage approach that imposes a constraint on the transition from the unreduced to the partially reduced incentive structure, and then a constraint on the transition from the partially reduced to the completely reduced incentive structure.

First, take the step from the unreduced to the partially reduced incentive structure of an agent. To obtain the partially reduced structure, it is

106

provisionally assumed that the agent pursues incentives relentlessly. The partially reduced structure comprises the incentives pursued if the agent pursues an incentive at every opportunity. Where there are multiple incentives away from an option, one incentive is selected for pursuit. Each option that generates an incentive in the unreduced structure generates exactly one incentive in the partially reduced structure. Among incentives to switch away from an option in the unreduced structure, the partially reduced structure retains the incentive pursued given that one is pursued.

What rules of rationality govern the transition from an agent's unreduced incentive structure to his partially reduced incentive structure? We propose the following constraint:

The partially reduced incentive structure is such that no incentive in it is insufficient if a replacement is sufficient in the structure that results from making the replacement.

The rule of rationality that requires the partially reduced incentive structure to comply with this constraint we call the *selection rule*. The selection rule concerns the choice of an incentive to be pursued from among an agent's incentives given an option. The rule says that sufficient incentives have priority over insufficient incentives. That is, incentives that initiate terminating paths have priority over those that do not. The selection rule concerns sufficiency with respect to the partially reduced incentive structure. It gives priority to incentives that are sufficient given minimal modifications of the structure. According to the selection rule, every unselected alternative to any incentive that is part of a nonterminating path for an agent in his partially reduced incentive structure is an incentive that does not initiate a terminating path for the agent if substituted in that incentive structure.

Figure 4.7 depicts the kind of case the selection rule prohibits. The option does not initiate a terminating path in the partially reduced incentive structure of the agent. It initiates a path, but the path does not terminate; it cycles back to the option. On the other hand, the agent has an unselected incentive away from the option that initiates a terminating path of incentives if substituted. The selection rule prevents the selection of an incentive that initiates a nonterminating path in the partially reduced incentive structure if there is an alternative incentive that initiates a terminating path when substituted in the partially reduced incentive structure.

To illustrate, consider the unreduced incentive structure in Figure 4.8. Since the path A, B terminates, whereas the path A, C, D, E, ... does not, a rational agent pursues the incentive from A to B instead of the incentive from A to C.

The selection rule is a *global* rule for selection of incentives. It applies to the whole process of selection, not to each selection one by one. It states that selection should favor incentives that are sufficient rather than insufficient

Unselected incentive

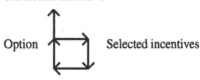

Option   Selected incentives

Figure 4.7   An unreduced incentive structure.

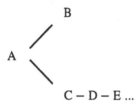

Figure 4.8   The selection rule at work.

in the structure generated. The constraint it imposes is that no selected incentive is insufficient given the global pattern of selection of incentives if an alternative incentive is sufficient given the same global pattern except for substitution of the alternative incentive.

To say that the selection rule is global as opposed to local is to say something about its input and output. The selection rule has as input the unreduced incentive structure of an agent. It has as output the partially reduced incentive structure of the agent. It imposes a constraint on the relationship between input and output. A local version of the selection rule has as input a node of the agent's unreduced incentive structure and has as output the corresponding node of his partially reduced incentive structure. It imposes a constraint on the relationship between the input and output nodes. It gives priority to sufficient incentives at a node. Because of the interdependency of pursuit and sufficiency of incentives, an agent may satisfy the local selection rule at each node of his unreduced structure but nonetheless fail to satisfy the global selection rule. After satisfying the local selection rule at a node, the status of incentives at previous nodes may change from sufficient to insufficient. In virtue of such changes, the global selection rule may be violated.

The global selection rule requires a type of coherence in the partially reduced incentive structure. The general principle of coherence for pursuit of incentives states that an agent must treat each incentive in a way that is rational given what he does with other incentives. The rationality of pursuit of incentives is thus, at least in part, a matter of conditional rationality. A pattern of pursuit of incentives is rational only if the treatment of each

108

incentive is rational given the pattern. The global selection rule results from applying the general principle in light of rules of rationality concerning sufficient incentives.

Now let us consider the transition from the partially reduced incentive structure to the completely reduced incentive structure. The completely reduced incentive structure is obtained by deleting incentives to switch strategy in the partially reduced incentive structure that the agent does not pursue. Incentives of the unreduced structure discarded by the partially reduced structure are not resurrected in the completely reduced structure. Rather, additional incentives are deleted. A previously deleted incentive is not pursued even though no alternative is pursued instead. Since not all the incentives of endless paths in the partially reduced structure are actually pursued, the deletion of unpursued incentives stops those paths. In an agent's partially reduced structure he has at most one incentive away from an option, and failing to pursue it amounts to stopping pursuit of incentives at that option.

Stopping pursuit of incentives is rational only if the incentive not pursued is insufficient in the partially reduced structure: that is, only if it does not initiate a terminating path of incentives. Since an incentive initiates just one closed path in the partially reduced incentive structure, an option's failure to initiate a terminating path in that structure is equivalent to the nontermination of the closed path it initiates, and this entails that the closed path is endless. We call the injunction to pursue sufficient incentives of the partially reduced structure the *stopping rule*. It requires pursuing incentives of the partially reduced incentive structure unless they are insufficient with respect to that structure. In other words, it imposes the following constraint:

The completely reduced incentive structure includes all sufficient incentives of the partially reduced incentive structure.

The argument for the stopping rule is straightforward. An agent's partially reduced incentive structure gives an agent's pattern of pursuit of incentives if he pursues incentives at every opportunity. This pattern determines whether pursuit of an incentive in the pattern is futile. Rationality requires pursuit of incentives where it is not futile, and so authorizes stopping pursuit of incentives only where pursuit is futile.

As other rules of rationality do, the stopping rule presumes that the input for it is rational, and so presumes that the agent's partially reduced incentive structure is rational. In case of irrational input, compensation for the irrationality may lead to deviation from the rule.

The stopping rule, as the selection rule, is a global rule. Its input and output are entire incentive structures. The rule's input is the partially reduced structure, and its output is the completely reduced structure. The rule puts a constraint on the relationship between input and output. Every sufficient incentive of the partially reduced structure is retained in the completely reduced structure. However, the local version of the rule is obeyed just in case the global version is. So the distinction between the rule's global and local versions is not important.

Our global rules of rational pursuit of incentives govern the transitions from unreduced to partially reduced structures, and from partially reduced to completely reduced structures. The rules explain why nonpursuit of an incentive is rational in terms of its insufficiency with respect to an appropriate incentive structure. To assess for sufficiency an incentive omitted from the partially reduced incentive structure, we restore all such omitted incentives to obtain the partially reduced incentive structure. And to assess for sufficiency an incentive omitted from the unreduced structure, we substitute it in the partially reduced incentive structure. In general, to assess an unpursued incentive for sufficiency, we see whether it starts a terminating path in the partially reduced incentive structure or has a rival that starts a terminating path in that incentive structure. This global approach to sufficiency of incentives handles the dependency of sufficiency on pursuit of incentives.[12]

To illustrate the various incentive structures and the rationality constraints on them, consider an ideal version of Matching Pennies in pure strategies. Suppose that Column pursues all incentives. Let us consider Row's incentives under that supposition. First, take Row's unreduced incentive structure. It incorporates all incentives pursued or not. His unreduced structure can be represented by the following endless path: H, T, H, T, ... Next, to obtain his partially reduced incentive structure, discard all incentives except those pursued if pursuit of incentives is relentless. In this case there are no incentives to discard. None of the incentives of the unreduced structure is deleted in the transition to the partially reduced structure since no nodes of the unreduced structure have multiple incentives. Moreover, the selection rule is satisfied automatically. Since there are no omitted incentives, no omitted incentive is sufficient when substituted for an insufficient incentive of the partially reduced structure. We assume that Row's pattern of relentless pursuit of incentives meets all other rationality constraints and is rational.

Finally, the completely reduced incentive structure is obtained by deleting unpursued incentives of the partially reduced structure. The deleted incentives are unpursued not because of rivals but because of realism. The stopping rule permits resisting an incentive only if it is insufficient in the partially reduced incentive structure, that is, only if pursuit of the incentive does not terminate in a tree in the partially reduced structure. In that case

[12] Given the central role of the partially reduced incentive structure and the reduction of trees to paths in the partially reduced incentive structure, termination in a tree, as opposed to termination in a subpath, is not crucial in our examples. We save further exploration of it for future work.

relentless pursuit of incentives leads to endless, futile pursuit of incentives. On the other hand, if in the partially reduced incentive structure, pursuit of an incentive away from an option terminates in a tree, then the incentive must be pursued; its pursuit is not futile.

Since it is impossible in our version of Matching Pennies for each agent to pursue incentives at every opportunity, some agent must abandon pursuit of an incentive. We suppose Row does. Imagine that Row stops pursuit of incentives at the first occurrence of T in the preceding path. Removing abandoned incentives yields his completely reduced incentive structure. Since the incentives after the first occurrence of T are insufficient in his partially reduced structure, his stopping pursuit of incentives at the first occurrence of T complies with the stopping rule.

To illustrate incentive structures and selection and stopping rules further, let us return to the example of Figure 4.5. In moving from the unreduced to the partially reduced incentive structure, we suppose that optimal incentives are selected. Then given R3 and hence (R3, C2), either the incentive to switch from R3 to R2 or the incentive to switch from R3 to R1 is selected. Neither of these two incentives initiates a terminating path in a partially reduced incentive structure containing it. So the selection rule allows either incentive to be pursued. We assume pursuit of the incentive to switch from R3 to R2. As a result, there is a terminating path away from R1 in the partially reduced incentive structure. It is R1, R3, R2, R3, ... This path terminates in the tree or subpath R3, R2, R3, ... The terminal cycle does not contain R1, is closed in the partially reduced incentive structure, and meets the minimality condition. Hence the incentive to switch from R1 to the tree beginning with R3 is sufficient. In consequence, the stopping rule requires pursuit of the incentive to switch from R1 to R3. We assume that the endless path R1, R3, R2, R3, ... is brought to a halt by disregarding the incentive to switch from R2 to R3. Stopping at R2 conforms with the stopping rule since R2 initiates a nonterminating cycle of incentives in the partially reduced incentive structure. Consequently, in the completely reduced incentive structure R1 initiates a terminating path: R1, R3, R2. According to the principle of self-support, R1 is therefore not a rational choice.

We obtain a different verdict about R1 if given R3 and hence (R3, C2) Row breaks the tie between R1 and R2 in favor of R1, and so pursues the incentive to switch from R3 to R1. Then the incentive to switch from R1 to R3 is not sufficient in the partially reduced incentive structure. It starts the path R1, R3, R1, ... This path does not terminate in the tree or subpath beginning with R3 since that subpath is a cycle containing R1. Hence the stopping rule permits disregarding the incentive to switch from R1 to R3. If that incentive is not pursued, then the completely reduced incentive structure has no path away from R1, and R1 is self-supporting – it may be a rational choice. Therefore the way in which given R3 Row breaks the tie between R1 and R2 affects the status of choosing R1.

Our characterizations of sufficient incentives and rational pursuit of incentives may seem circular. We say that an incentive is sufficient if and only if it does not initiate a terminating path in the completely reduced incentive structure. But we assume that the characterization is applied to an ideal agent who pursues incentives rationally. We also say that rational pursuit of incentives meets the selection and stopping rules. These rules require giving priority to pursuit of sufficient incentives. So the direction of rational pursuit of incentives depends on the sufficiency of incentives, and the sufficiency of incentives depends on the direction of rational pursuit of incentives.

To put the worry in another light, the account of sufficient incentives seems to generate a circle in which rational pursuit of incentives is taken as the grounds of rational pursuit of incentives. For suppose that an incentive is not pursued. Then it is not part of the completely reduced incentive structure. So it does not start a terminating path in the completely reduced structure. Thus according to the account of sufficient incentives, the incentive is insufficient, and so rationally not pursued. However, we do not want to say that an incentive's exclusion from the completely reduced structure by itself shows that the incentive is insufficient. That would be circular justification of the incentive's nonpursuit.

The appearance of circularity is dispelled by attending to sufficiency's relativity to an incentive structure. Although the sufficiency of incentives in the completely reduced incentive structure depends on the direction of rational pursuit of incentives, the direction of rational pursuit of incentives depends on the sufficiency of incentives in the partially reduced incentive structure. Take an incentive that is insufficient in the partially reduced incentive structure. The selection rule requires that rival incentives be insufficient in that structure. Next, take an incentive that is sufficient in the partially reduced incentive structure. The stopping rule entails that it is sufficient, period. There is no circular dependence of the selection and stopping rules on the sufficiency of incentives in the completely reduced structure. Although an incentive's absence from the completely reduced structure of an ideal agent rests on the assumption that its nonpursuit is rational, we do not claim that an incentive missing from the completely reduced structure is rationally not pursued just because it is not pursued. We do not justify an incentive's nonpursuit simply by pointing out that the incentive is not part of the completely reduced structure. We explain the rationality of its nonpursuit in terms of global rules of rational pursuit of incentives applied to the unreduced and partially reduced incentive structures.

Next, let me add a gloss on the role of the selection and stopping rules. Our account of rational pursuit of incentives presumes compliance with the selection and stopping rules. But the selection and stopping rules are not meant as a full account of rational pursuit of incentives. There may be additional rules governing rational pursuit of incentives. We assume that the ideal agents in our examples satisfy all rules for rational pursuit of incentives, including any that goes beyond the selection and stopping rules.

Some additional rules for rational pursuit of incentives may be related to reasons that are independent of incentive structures. We assume that ideal agents follow those rules. So, for instance, in examples in which an agent abandons pursuit of an incentive that does not initiate a terminating path in his partially reduced incentive structure, we assume that there are no reasons independent of the partially reduced incentive structure that make the incentive sufficient. As noted earlier in the section, we treat only games in which termination of paths settles sufficiency of incentives.

Some additional rules for rational pursuit of incentives may use incentive structures to constrain further selection of incentives and stopping pursuit of incentives. For instance, at each node in an agent's unreduced incentive structure, rationality may require selection of an optimal incentive, that is, an incentive that is optimal among the incentives at the node. We leave this particular issue open. Although optimality is attractive, some considerations suggest that it is not required. Strategy may favor pursuit of nonoptimal incentives in some circumstances because of the responses of other agents to it. Also, a rule requiring optimality may conflict with the selection rule. The selection rule advises pursuing a unique sufficient incentive. But what if that incentive is nonoptimal? Wouldn't sufficiency take precedence over optimality? I do not try to settle the issue. I do not try to determine whether optimality conflicts with sufficiency, and, if so, whether rationality gives precedence to sufficiency. I evade the issue by using examples in which agents pursue optimal incentives without violating the selection rule. I leave it to future studies to determine what requirement of optimality, if any, is warranted.[13]

There may also be a rule that requires stopping after pursuit of an incentive unless the grounds of preference change. This would prevent cycles among tying options, and gradual ascent up a stable preference ranking of options, as opposed to immediate movement to the top of the preference ranking. In examples we assume that agents comply with this rule, but we leave the issue of mandatory compliance open.[14]

Additional principles concerning rational pursuit of incentives are suggested by reflection on the consequences of compliance with the selection

---

[13] For more on this issue, however, see the note in Section 4.5 on the mixed extension of Matching Pennies. Although examples generally assume pursuit of optimal incentives, to entertain objections, the games of Figures 4.9 and 8.9 allow pursuit of nonoptimal incentives.

[14] Although I maintain neutrality about indecisive switching, I entertain criticism that takes it to be rational in the game of Figure 8.9.

rule. In some cases pursuit of an insufficient incentive leads to a better outcome than pursuit of a sufficient incentive. Newcomb's problem teaches us to accept cases in which rationality is penalized (as Section 4.2 mentions). So we do not take such cases as an objection to the selection rule. However, since an agent's pursuit of incentives affects the sufficiency of his incentives, an agent may have some motivation to alter the incentives he pursues in order to alter the incentives that are sufficient and make them more rewarding. An alteration in the incentives that are sufficient may alter the options that are rational, and thus make more attractive options become rational options. An agent's pursuit of incentives is a matter of his dispositions to choose with respect to subsets of options, and an agent brings these dispositions with him to his decision problem. At the time of the problem, it is too late for him to alter these dispositions. There is no backward causation from choice to disposition. So strictly speaking, in the course of a decision problem he cannot alter his pursuit of incentives to make it more rewarding. But prior to the decision problem, an agent can cultivate dispositions that generate a new pattern of pursuit of incentives with respect to which the options that are rational may be more rewarding. There may be interesting principles of rationality governing such preparations for decision problems; however, exploration of this area would take us too far afield.

In general, we assume that ideal agents pursue incentives in a rational way but do not fully explain what this entails. It is useful to give a partial characterization of rational pursuit of incentives to flesh out and clarify examples; the selection and stopping rules do this. But a complete characterization of rational pursuit of incentives is beyond this book's scope. Our main proposals concerning equilibrium are independent of issues concerning rational pursuit of incentives. In particular, strategic equilibria exist with respect to any pattern of pursuit of incentives, as Chapter 5 shows. Consequently, we can put aside difficult issues about rational pursuit of incentives. We need keep in mind only that profiles that are strategic equilibria with respect to a certain pattern of pursuit of incentives may fail to be solutions because the pattern of pursuit of incentives is irrational.

4.5. COMPARISON OF DECISION PRINCIPLES

The next chapter uses the principle of self-support to formulate a new equilibrium standard for solutions to games. But the principle is a decision principle, intended for all decision problems, not just those that arise in games. This section more firmly establishes the decision principle's credentials. I compare the principle with other, familiar decision principles: the principle of dominance, principle alpha, the principle of ratification, and the expected utility principle. I look for cases in which the principle of self-support conflicts with one of these other principles, and in such cases defend the principle of self-support. In comparisons I assume that agents are ideal in the decision-theoretic sense, and so cognitively unlimited and in rational states of mind but capable of choosing irrationally. I also assume that reasons besides incentives are negligible so that strategic self-support is the only kind that matters.

4.5.1. Common Decision Principles

To set the stage for comparisons of the principle of self-support with the other decision principles, I present the other principles and discuss their relations to each other. I begin with the principle to maximize expected utility. There are several versions of the principle. One version simply says to pick an option of maximum expected utility (MEU). Formulated this way the principle is meant to express a *sufficient* (S) condition for rational choice. Call this principle MEU-S. An option of maximum expected utility exists only if every option has an expected utility. MEU-S recognizes that there are cases in which no option has maximum expected utility because some options lack expected utilities – crucial beliefs and desires are not quantitative. It states only that in cases in which an option of maximum expected utility exists, picking such an option suffices for choosing rationally. In looking for conflict with the principle of self-support, a principle that expresses a necessary condition of rational choice, we look for cases in which an option that maximizes expected utility is not self-supporting.

Another version of the expected utility principle advises picking an option of maximum expected utility if every option (of an appropriate partition) has an expected utility. This version of the principle is meant to express a *necessary* (N) condition of rational choice; call the principle MEU-N. It is not intended to express a sufficient condition of rational choice. It foresees cases in which no option has maximum expected utility because crucial beliefs and desires are nonquantitative. It does not say that if some option lacks an expected utility, anything goes. It allows for additional principles of rationality to govern decision problems in which some options do not have expected utilities. It even allows for additional principles of rationality for selecting among multiple options with maximum expected utility. Conflict between MEU-N and the principle of self-support arises when it is impossible to satisfy both principles because no option that satisfies MEU-N also satisfies the principle of self-support. This happens if some options have maximum expected utility, but none of those options is self-supporting.

A generalization of the principle to maximize expected utility says to choose so that the option chosen maximizes expected utility under some *quantization* of belief and desire. It advances a *necessary and sufficient* (NS)

115

condition of rational choice designed to accommodate cases where beliefs and desires are not quantitative. Call this principle MEU-NS. It conflicts with the principle of self-support wherever the options it proscribes include all the self-supporting options.

Section 3.2.2 introduced the principle of ratification. We review the principle here and elaborate a few of its features. A ratifiable option is an option that has maximum expected utility on the assumption that it is adopted. The principle of ratification says to pick a ratifiable option. The principle is a version of the principle to maximize expected utility where maximization of expected utility is relative to the assumption of an option. The relativization makes sense since an option's expected utility, as an assessment of its choiceworthiness, should take account of relevant information carried by the assumption that the option is adopted.

The principle of ratification is a precursor of the principle of self-support. Assuming that expected utility comparisons conditional upon an option exist and that preferences conditional upon the option agree with those comparisons, the principle of ratification is the same as the principle of adoption of an incentive-proof option. In light of cases without incentive-proof options, we revise the principle of incentive-proofness, or ratification, to obtain the principle of self-support. We accommodate the intuition that there is a rational choice in every decision problem by not requiring pursuit of incentives to switch options if switching options is futile.

In cases in which there are no ratifiable options, some defenders of the principle of ratification say that no choice is rational. They take ratification as necessary for rational choice. In some cases in which there are several ratifiable options, intuition indicates that not all are rational choices. So ratification is generally taken as only a necessary condition of rational choice. Supplementary principles are envisioned for choice among ratifiable options. Taking the principle of ratification to express a necessary condition of rational choice, and assuming that where ratifiable options exist some choices are rational, the principle conflicts with the principle of self-support if meeting each principle is possible, but meeting both is not possible.

The principle of dominance that I consider advises not picking an option that is strictly dominated by another option if the ordering of options by strict domination is finite. It expresses the view that when it is possible, avoiding strict domination is necessary for rational choice. I formulate the principle in terms of strict domination since ordinary domination is not sufficient to ground a preference. An option *dominates* another option in the ordinary way if and only if under some partition of possible states of the world that are causally independent of the options, in every state the agent prefers the first option to the second option or is indifferent between them and in some state prefers the first option to the second option. It may be that the states in which

116

the first option is preferred are certain not to occur. So all that follows is that the second option is not preferable to the first. It does not follow that the first option is preferable to the second. On the other hand, an option *strictly dominates* another if and only if under some partition of possible states of the world that are causally independent of the options, the first option is preferred to the second option in all states. Strict domination entails that the first option is preferable to the second. Consequently, the principle of strict dominance expresses a plausible necessary condition of rational choice.[15]

We assume that for ideal agents, strict dominance is transitive. This follows from the transitivity of preference if the same partition of possible states of the world is used for all options. When different partitions are used for different pairs of options, we assume that some rule of coherence for preference prohibits the possibility of a partition with respect to which A strictly dominates B, and a partition with respect to which B strictly dominates C, but no partition with respect to which A strictly dominates C. The rule may be that strict domination with respect to one partition entails strict domination with respect to every coarser partition. Then to show transitivity of strict domination, it suffices to construct a partition for A and C coarser than the ones for A and B, and for B and C.

Principle alpha concerns coherence of choices in two decision problems, decision problems that arise before and after a contraction in the set of options available. The principle says that an option that is rational given one set of options is also rational given a subset that contains it. In other words, it states that if an option A is a rational choice in one decision problem, then it is a rational choice in another decision problem in which the options are the same as in the first problem except for the omission of options other than A – that is, the options in the second problem are a subset of the options in the first problem, and A is an option in both problems. Notice that the principle does not say that the option chosen must be the same in both decision problems. It states only that if a certain option is a rational choice in the first problem, it is a rational choice in the second. In decision problems in which there are several rational options, not all can be chosen, and principle alpha allows for this.[16]

---

[15] For more on the principle of dominance and the restriction concerning causal independence, see Gibbard and Harper (1978, Sec. 8).

[16] Principle alpha is usually advanced with a companion principle beta concerning expansion of the set of options available. Principle beta states that if two options are rational choices in a decision problem, then they are both rational choices, or both not rational choices, in any decision problem with options that form a superset of the options in the first decision problem. We put aside principle beta because it raises the same issues as principle alpha. Sen (1970: 17) presents principles alpha and beta as properties of a rational choice function C from a set of options S in a base set X to a subset of S. He says C has property alpha if and only if, for every option x, $x \in S_1 \subset S_2 \rightarrow [x \in C(S_2) \rightarrow x \in C(S_1)]$. He says C has the companion property beta if and only if, for all options x and y, $[x, y \in C(S_1) \& S_1 \subset S_2] \rightarrow [x \in C(S_2) \leftrightarrow y \in C(S_2)]$.

Principle alpha expresses a necessary condition for rational choices. It is subject to the proviso that the grounds of preference for the options common to the two decision problems are the same in both problems. That is, the reasons for forming preferences among those options are the same in both problems. One motivates the principle by assuming that there is a preference ranking of actions that underlies the preference ranking of options or *feasible* actions, and that the preference ranking of actions does not change as the set of options changes. Or more generally, one assumes that the grounds of preference among actions are stable, and if they produce a preference ranking of actions, produce a stable one, a ranking that does not depend on the actions available as options. Violations of principle alpha are clearly permissible in cases in which the availability of options affects preferences between options by changing the grounds of preference, as noted by Luce and Raiffa (1957: 288), Eells and Harper (1991), and Vallentyne (1991: 316). This happens, for instance, in cases in which removal of an option changes the preference ranking of remaining options by, say, providing information about those options. We formulate the proviso so that it says that the grounds of preference are the same for the options common to the two decision problems, instead of saying that those options have the same preference ranking in both problems, in order to allow for cases in which a preference ranking of the common options is nonexistent or incomplete.

I take all the familiar decision principles reviewed previously to be subordinate to some version of the principle to maximize expected utility. MEU-N says to pick an option of maximum expected utility if every option has an expected utility. The principle of ratification is a generalization of MEU-N for cases in which assumption of an option carries pertinent information. The principle of ratification is not an additional principle to be met along with MEU-N, but a generalization of MEU-N for cases in which it is anticipated that a decision will itself provide information about the possible states of the world used to compute expected utilities. MEU-N follows from the principle of ratification when the assumption of an option does not carry relevant information.

The principle of dominance follows from MEU-NS, which specifies choosing so that the option chosen maximizes expected utility under some quantization of belief and desire. It is a version of the principle to maximize expected utility for cases in which some options lack expected utilities. If every option has an expected utility, the principle of dominance follows from MEU-N; dominance supplements MEU-N only in cases in which some options lack an expected utility. Principle alpha follows from MEU-NS applied to pairs of decision problems with the same grounds of preference for the common options. If an option maximizes expected utility under some quantization of the larger decision problem, it maximizes expected utility under some quantization of the smaller decision problem.

The versions of the principle to maximize expected utility are themselves subordinate to some more basic decision principles, for instance, the principle to maximize utility. Expected utility is just utility for gambles. The expected utility principles lay down ways to pursue the objective of maximum utility when there are obstacles such as uncertainty and indeterminate probabilities. In addition, the principle to maximize utility is subordinate to the principle to pursue incentives: Do not adopt an option if another is preferred. Maximizing utility is the result of pursuing all incentives in quantitative cases. Also, making the principle to pursue incentives relative to an option, we obtain the principle of incentive-proofness: Do not adopt an option if given it there is another preferred. It yields the principle of ratification in the quantitative cases we generally assume.

To begin comparing the principle of self-support with the other familiar decision principles, let us consider in a general way how the principle of self-support and the other principles apply to various types of decision problem.

Although concrete decision problems with ideal agents involve an infinite number of options, all the propositions the agent may decide upon, *representations* of decision problems typically list a finite partition of options. It is assumed that the partition of options is *appropriate*, that is, that it forms an appropriate set of options to consider and so includes the main contenders. We generally ignore the distinction between decision problems and their representations. When we speak of a decision problem, we usually have in mind a representation of a decision problem that provides an appropriate partition of options. (Compare Section 1.3 on the distinction between games and their representations.)

In a standard decision problem the preference ranking of options is complete, has a top rank, and is stable. That is, (1) each pair of options is compared, and the agent prefers one to the other or is indifferent between the two; (2) the preference ranking of options has a top rank; and (3) the assumption of an option does not alter comparisons of options; conditional comparisons are the same as nonconditional ones.

By a top rank of a preference ranking of options, I mean a rank such that no option has a higher rank. A preference ranking of options has a top rank if it ranks a finite number of options and so has a finite number of ranks. The second standard feature of decision problems is formulated under the assumption that the first standard feature obtains. We also present a general formulation that accommodates cases in which the first standard feature does not obtain, and the preference ranking of options is incomplete. In that case a finite preference ranking may have several top ranks. The general version of (2) states that the preference ranking has no infinite subranking. Where (1) is not met, we substitute the general version of (2).

In the standard case the principle of self-support eliminates all options except those at the top of the preference ranking. For each option not at the top, there is a path of incentives away from the option that terminates with a top option. All of the top options meet the principle of self-support since there are no incentives to switch from them to other options, and so no terminating paths of incentives away from them. In this case meeting the principle of self-support is necessary and sufficient for meeting MEU-NS, the generalization of the principle to maximize expected utility for cases in which some options lack expected utilities. If all options have expected utilities, meeting the principle of self-support is also necessary and sufficient for meeting MEU-N. All the options that satisfy the principle of self-support are rational choices, assuming that reasons besides incentives are negligible.

For each of the three features of standard decision problems, there is a nonstandard case in which just that feature is missing. First, suppose that the preference ranking is incomplete but is stable and has no infinite subranking. In this case the options meeting the principle of self-support may not be compared. For instance, the options forming the termini of paths of incentives may not be compared. Although options not meeting the principle, that is, options before terminal options, are not rational choices, there is no guarantee that all options satisfying the principle, that is, terminal options, are rational choices. Some terminal options may fail to be rational choices because, in virtue of reasons besides incentives, they are less choiceworthy than other terminal options.

Also, in this case meeting the principle of self-support is not sufficient for maximizing expected utility. Whereas there are self-supporting options in every decision problem, in some problems there are no options that have maximum expected utility. Being an option of maximum expected utility entails (for ideal agents) being preferred to all options of nonmaximum expected utility. If the preference ranking is incomplete, there may be no option that satisfies this condition.

The principle of self-support, however, agrees with MEU-N, the principle to maximize expected utility if all options have an expected utility. Suppose that there is an option of maximum expected utility. Then it is self-supporting. There is no terminating path away from it. Also, if an option is rejected by MEU-N because another option has greater expected utility, then the option is rejected by the principle of self-support. There is a path from the first option to the second. The path never returns to the first option and terminates with the second option or with some other option preferred to both the first and the second. Moreover, if an option is rejected by the principle of dominance, it is rejected by the principle of self-support. There is a path

120

away from the dominated option to the dominating option that never returns to the dominated option. It terminates with the dominating option or with some other option preferred to both the dominated and the dominating option.

Next, suppose that we depart from the standard case by allowing the preference ranking to have no top rank, although it is complete and stable. In this case for each option there is another preferred. No option initiates a terminating path of incentives. No path terminates in an option. And none terminates in a tree either; no tree meets the minimality condition since given stable preferences no option is repeated in a path, and thus in a tree. As a result, the principle of self-support does not eliminate any option. But it does not follow that all options are rational choices. Some may be irrational because of the structure of the preference ranking and their position in it. For instance, the preference ranking may be bounded above and below. Then options near the lower bound are irrational. Every choice violates MEU-N if expected utility comparisons follow preferences. And if preferences follow strict dominance, every choice also violates the principle of dominance, putting aside the principle's proviso that the order of options by strict dominance is finite. Principle alpha is violated if, by the principle of satisficing, say, some option is rational even though another option is preferred. If all options except those two are removed, the second clearly becomes the unique rational choice, contrary to principle alpha. I hold that some choice is rational in each decision problem of this second nonstandard type, and so reject MEU-N and alpha as general decision principles, and reject dominance without the proviso.[17]

Let me dispose of one objection to the principle of self-support in such cases. Someone may object that there are cases of this type in which initiation of a terminating path in a decision problem's completely reduced incentive structure does not disqualify a strategy as a rational choice. Consider a realization of the decision problem of choosing a natural number that gives your income in dollars. Suppose that in the partially reduced incentive structure, for any number, there is an incentive to switch to the successor of that number. Other incentives have been discarded; they are not pursued. In the completely reduced incentive structure all nonterminating paths of the partially reduced incentive structure halt. That is, in the completely reduced incentive structure a path away from a number halts at some number. It may seem that an option that initiates a terminating path in the completely reduced incentive structure is not thereby irrational. An option that initiates a terminating path may not seem to be irrational since failure to pursue

---

[17] "Satisficing" is commonly proposed when the preference ranking of options is infinite. Although satisfactoriness is vague, it is in fact a necessary condition of rational choice. Still, it is not a sufficient condition of rational choice. In particular, it ignores self-support. A satisfactory but self-defeating option is not rational.

incentives past the path's terminus may not seem to make prior incentives, including the first incentive away from the option, sufficient. For example, consider an agent whose path away from $T ends at $M because he does not pursue incentives to switch to incomes higher than $M. He may seem not to have a sufficient incentive to switch from $T to $T + $1, just as he does not have a sufficient incentive to switch from $M to $M + 1. But the first incentive is sufficient since its pursuit is not futile. Its pursuit eventually leads to $M, where pursuit of incentives stops. The place where pursuit of incentives stops may be somewhat arbitrary, but once a stopping place is fixed, rationality requires pursuing incentives up to it.[18]

Finally, for the third nonstandard case, suppose that the preference ranking is unstable but is complete and has a top rank. The preference ranking's instability entails that it changes under the assumption that an option is adopted. To illustrate, suppose that an agent is indifferent between options A and B, but prefers A to B assuming that B is adopted, and prefers B to A assuming that A is adopted. And suppose that paths of incentives away from other options terminate in the cycle from A to B and back to A. The principle of self-support eliminates options preceding such terminal cycles, as well as options preceding terminal options. So it eliminates all options besides A and B. It also eliminates either A or B depending on which generates an unpursued incentive; the remainder is the rational choice. This result is contrary to the principle of ratification. The principle of ratification states that a rational choice maximizes expected utility on the assumption that it is adopted. But neither A nor B meets this condition.

The principle of self-support does not directly conflict with the principle of ratification. There are no cases in which it is possible to satisfy each principle but impossible to satisfy both. Whenever an option is ratifiable, it is also self-supporting. If an option is ratifiable, there are no incentives to switch from it. And if there are no incentives to switch from it, the option does not initiate a terminating path of incentives and so is self-supporting. However, self-supporting options are more prevalent than ratifiable options. There are cases in which no option is ratifiable but some options are self-supporting. These cases may involve a preference ranking with no top, as in the decision problem of choosing one's income, or unstable preferences, as in Matching Pennies. The example of the previous paragraph involves unstable preferences. Given the intuition that in every decision problem there is a rational choice, these cases make it implausible that ratifiability is a necessary condition for a rational choice. That is why we propose

[18] Cf. Bratman (1987: 22–3) on Buridan's ass. Bratman contends that after breaking a tie between plans, an agent is committed to the plan chosen. It would be irrational to switch plans without reason. This is another case in which settling an arbitrary matter has implications for rational choice.

self-support as a substitute. The attainability of rationality favors the principle of self-support over the principle of ratification.

4.5.2. *Conflict between Decision Principles*

In the third type of nonstandard decision problem, the principle of self-support conflicts with MEU-S, MEU-N, dominance, and alpha. This section displays the conflict and defends the principle of self-support. To make conflict with the principle of self-support easier to generate, we allow nonoptimal pursuit of incentives by ideal agents. This is an exception to our general policy of neutrality on the issue. We make the exception as an allowance to criticism of the principle of self-support. Wherever nonoptimal pursuit of incentives plays a role in examples, we draw attention to it. After the section closes, we return to our policy of neutrality.

Some of the examples used to bring out a conflict of decision principles involve games. But we treat each game only as a context for an individual decision problem. We take agents to be ideal only in the decision-theoretic sense and so understand them to be capable of an irrational choice. We do not attempt to find a solution to the game. In each game we are interested in an agent's rational choices. A rational choice is not the same as a strategy that is part of a solution to the game. A solution is a profile of strategies such that each is rational *given the profile*. In this section we look for strategies in games that are unconditionally rational. In other words, we look for strategies that are rational given actual circumstances, including the actual choices of agents.

Let us show that the principle of self-support conflicts with MEU-S and MEU-N. To begin, we have to specify the information with respect to which expected utility is taken. I suppose it is the information an agent has just before he makes a choice and includes foreknowledge of his choice. This is appropriate for ideal agents. Also bear in mind that self-support is assessed with respect to the agent's completely reduced incentive structure, that is, the incentive structure that results after discarding incentives not pursued. An option is self-supporting if and only if it does not initiate a terminating path in that structure.

The principle of self-support conflicts with MEU-S, the principle that advances maximization of expected utility as a sufficient condition of rational choice. There are cases in which an option of maximum expected utility is not self-supporting. For example, take an ideal version of the mixed extension of Matching Pennies in which Row adopts a 50–50 probability mixture of H and T and knows that Column also adopts a 50–50 probability mixture of H and T. Row's adopting any other mixture, say, a 30–70 mixture of H and T, also maximizes expected utility. But if Row adopts the 30–70

123

mixture, we suppose he knows that Column adopts H, and then instead of the 30–70 mixture he prefers the 50–50 mixture. If he switches back to the 50–50 mixture he has no incentive to switch from it. So there is a terminating path away from the 30–70 mixture. We can suppose it is a terminating path of Row's completely reduced incentive structure. Then the 30–70 mixture is not self-supporting.[19]

The principle of self-support also conflicts with MEU-N, the principle that advances maximization of expected utility if every option has an expected utility as a necessary condition of rational choice. For example, take an ideal version of Matching Pennies in which agents pursue their incentives except that Row, the matcher, rationally declines to pursue his incentive to switch from H to T. Row knows he will adopt H. This strategy does not maximize expected utility given his foreknowledge that Column responds with T. It is, however, Row's only self-supporting choice.

The cases in which the principle of self-support conflicts with expected utility principles are not damaging to the principle of self-support. They bring out the need to revise expected utility principles for cases in which the assumption that an option is adopted carries relevant information. The mixed extension of Matching Pennies shows that maximizing expected utility in a nonconditional way is not sufficient for rational choice. And our ideal version of Matching Pennies shows that it is not necessary for rational choice.

The principle of self-support also conflicts with the principle of dominance and principle alpha. There are cases in which the only options that are self-supporting are strictly dominated, or violate alpha.

In an unreduced incentive structure, in which all incentives to switch appear, those pursued and those not pursued alike, there is no conflict between the principle of dominance and the principle of self-support, provided that the preference ranking of options has no infinite subranking. In every case some options that are self-supporting are not strictly dominated. To see this, first notice that since strict dominance is transitive, it cannot turn out that every option that is self-supporting is strictly dominated by some option that is self-supporting. Among the self-supporting options, there must be a top option in their ordering by strict domination given that preference subrankings

---

[19] In this case the selection rule seems to require that a sufficient incentive be pursued instead of an optimal incentive. The selection rule says to favor sufficient incentives over insufficient ones in the move from the unreduced to the partially reduced incentive structure. Row's incentive to switch from the 30–70 mixture to the 50–50 mixture is a sufficient incentive, as we saw earlier. Row's optimal incentive away from the 30–70 mixture is the incentive to switch to H. But this incentive seems to be insufficient. It initiates a path to H, then back to the 30–70 mixture. Hence the selection rule seems to require pursuit of the sufficient incentive to switch to the 50–50 mixture over pursuit of the optimal incentive to switch to H. But this analysis is not right. All it takes for an incentive to be sufficient is initiation of one terminating path. The incentive to switch from 30–70 to H also initiates a path from H to 50–50, and this path terminates with 50–50. So the optimal incentive is a sufficient incentive too.

$$1, 2 \qquad 1, 1$$

$$2, 2 \qquad 2, 1$$

$$3, 1 \qquad 0, 2$$

Figure 4.9   Dominance versus self-support.

are finite, even if the ordering is incomplete. So let us consider an option A that is self-supporting and is not strictly dominated by any option that is self-supporting. It follows that A is not strictly dominated by any option whatsoever. For suppose that A is strictly dominated by a self-defeating option B. B initiates a terminating path. Because B strictly dominates A, there is an incentive to switch from A to B. Hence A also initiates a terminating path. So A is not self-supporting, contrary to assumption.[20]

Conflict between the principles of self-support and dominance may arise, however, in the completely reduced incentive structure for an agent, at least if we allow pursuit of nonoptimal incentives in the partially reduced incentive structure. This happens in an ideal normal-form game in pure strategies depicted by Figure 4.9 (a variant of the game of Figure 4.5). Suppose that given relentless pursuit of incentives, Row pursues incentives to switch from R1 to R3 and from R3 to R1, and Column pursues incentives to switch from C1 to C2 and from C2 to C1. Then the agents' pursuit of incentives is endless; they move in a cycle around the four corners of the matrix. Row's pursuit of the incentive from R3 to R1 instead of the incentive from R3 to R2 is nonoptimal but may seem not irrational in view of the entire incentive structure. Because Row's pursuit of incentives is endless if each agent pursues incentives relentlessly, Row does not actually pursue his incentive to switch from R1 to R3. In fact, his pursuit of incentives stops at (R1, C1). Row resists pursuit of his incentive to switch to R3, and given that Row's response to C1 is R1, Column has no incentive to switch. Each agent stops pursuit of incentives at (R1, C1). In his completely reduced incentive structure Row has terminating paths away from R2 and R3. The path from R2 is R2, R3, R1. The path from R3 is R3, R1. There is no path away from R1, however, since the incentives from R1 to R2 and from R1 to R3 are discarded. R1 is self-supporting despite the terminating path away from R2 because the incentive to switch from R1 to R2 is discarded

[20] Rabinowicz (1989: 630) presents a case in which some principles concerning self-supporting options conflict with the principle of dominance. But the case does not involve conflict between the principle of dominance and our principle of self-support, since one of the self-supporting options in his case is undominated. It involves conflict between the principle of dominance and a principle for choosing among self-supporting options advanced in Weirich (1988, Sec. 5).

125

in the completely reduced incentive structure. R1 is Row's one and only self-supporting option. Yet R1 is dominated by R2.

I reject the principle of dominance in this case. It is rational for Row to adopt R1 although that strategy is strictly dominated. It is rational for Row to resist his incentive to switch from R1 to R3. His resistance ends futile pursuit of incentives, and abandoning futile pursuit of incentives is rational. Self-support is attained if R1 is chosen, and attaining self-support is more important than avoiding strict domination.

There are three rationales for the principle of dominance. Each is undermined by attention to cases in which it is futile to pursue incentives.

1 The principle of dominance is supported chiefly by the principle to reject an option if another option is certain to result in more utility. But the latter principle has exceptions. In cases in which every option is such that another option is certain to result in more utility, the principle must be violated, and some option that violates it is rational. Cases of this type arise when the preference ranking of options is complete but has no top.

2 The principle of dominance is also supported by MEU-NS, the principle that advances maximization of expected utility under some quantization of belief and desire as a necessary and sufficient condition of rational choice. MEU-NS prescribes a way to pursue the goal of maximum utility in the face of obstacles such as nonquantitative belief and desire. But in the game of Figure 4.9 pursuing maximum utility is futile. No option has maximum utility on the assumption that it is adopted. Pursuit of maximum utility is endless. Rationality does not require an endless, futile pursuit of maximum utility. This observation undermines MEU-NS.

3 Another rationale for the principle of dominance is pursuit of incentives, not maximum utility. Complying with the principle of dominance is meeting the standard of nondominance, and the standard of nondominance is just the standard of incentive-proofness applied to cases in which strict dominance ensures the existence of an incentive. But incentive-proofness is not a necessary condition of rational choice, not when pursuit of incentives is futile. An incentive to switch to a strictly dominating option is just like any other incentive. If it initiates a terminating path, it is sufficient; otherwise not. An incentive arising from strict domination has no more force than other incentives. It may exist where probabilities for states of the world are unavailable. But this does not give it extra force. Not all incentives to switch are sufficient reasons to switch. Futile incentives are not sufficient reasons, and hence the futile incentives generated by strict domination are not sufficient reasons.

To obtain an acceptable principle of dominance, we must revise the principle so that it does not require futile pursuit of incentives. One way to do this is to let the principle's objective be the expression of a sufficient condition of conditional preference. Let it say that if one option B strictly dominates another option A on the assumption that A is adopted, then B should be preferred to A given A. This principle follows from a principle of ratification for preference generalized to accommodate cases in which belief and desire are not quantitative. That principle says to prefer B to A if, under some quantization of belief and desire, B has greater expected utility than A on the assumption that A is adopted. It entails the revised principle of dominance. For if on the assumption that A is adopted B is preferred to A in all the cases of an appropriate partition, then, under some quantization of belief and desire, B has higher expected utility than A on the assumption that A is adopted.

If we insist on a principle of dominance that expresses a necessary condition of rational choice, then the definition of strict dominance can be restricted to cases in which the assumption of an option does not carry information relevant to the comparison of options. The motivation for the restriction is similar to the motivation for assessment of strict domination with respect to a partition of cases that are not causally influenced by options. Just as causal influence upsets the basis of comparison of options, the assumption that an option is adopted upsets the basis of comparison of options if the assumption of the option carries information relevant to the comparison of options.

Finally, let the principle of self-support confront principle alpha. Principle alpha states that an option that is a rational choice in one decision problem is a rational choice in another decision problem obtained from the first by deletion of other options, provided the grounds of preference among remaining options do not change. Principle alpha conflicts with the principle of self-support because deletion of options may change the status of remaining options from self-defeating to self-supporting. A new self-supporter may upstage an option favored in the original decision problem. For example, in the game of Figure 4.9, R1 is Row's only self-supporting option, and so is his rational choice according to the principle of self-support. If R3 is deleted, we obtain the game of Figure 4.10.

In this game Row has an incentive to switch from R1 to R2 and no incentive to switch away from R2. If Row pursues incentives rationality, he

1, 2   1, 1

2, 2   2, 1

Figure 4.10   Alpha versus self-support.

127

pursues the incentive to switch from R1 to R2. So there is a terminating path away from R1 in his completely reduced incentive structure. R1 is not self-supporting. On the other hand, R2 is self-supporting. There is no incentive to switch from it, and hence no terminating path of incentives away from it in Row's completely reduced incentive structure. Since R2 is Row's only self-supporting option in the trimmed game, it is his only rational choice according to the principle of self-support.

Principle alpha yields a contrary result. In the original problem Row chooses from R1, R2, and R3. In the trimmed problem he chooses from R1 and R2. The removal of R3 does not bear on preference relations between the remaining options R1 and R2. The grounds of preference are belief and desire. In both the original and trimmed problems Row knows that Column adopts C1. In both the original and trimmed problems Row knows that the consequence of R1 is an outcome of utility 1 and the consequence of R2 is an outcome of utility 2. There is no change in the grounds of preference for R1 and R2. In fact, there is no change in the preference ranking of options. In both the original and the trimmed problem Row actually prefers R2 to R1. Given that R1 is a rational choice in the original problem, principle alpha says it is a rational choice in the trimmed problem. In our example it is possible to satisfy principle alpha, and possible to satisfy the principle of self-support, but it is impossible to satisfy both principles.

In cases of conflict such as the foregoing, we side with the principle of self-support. We reject principle alpha for the same reasons we reject MEU-NS, the expected utility principle on which principle alpha rests. There are cases in which pursuit of incentives is futile, and rationality does not require futile pursuit of incentives. In the example, removal of R3 does not change the grounds of *preference* for R1 and R2 but does change the grounds of their *choiceworthiness*. Preference is only one of the grounds of choiceworthiness; self-support is another. The elimination of R3 changes the grounds of choiceworthiness for R1 and R2 by changing the option that is self-supporting.

The grounds of preference between R1 and R2 do not include information about the futility of pursuit of incentives. In the original problem given relentless pursuit of incentives, Row pursues the incentive to switch from R1 to R3. He then has an incentive to switch from R3 to R1, which he pursues back to R1. Pursuit of the incentive to switch from R1 to R3 is futile. In the second problem, given relentless pursuit of incentives, he pursues the incentive to switch from R1 to R2. Although he may still have an incentive to switch from R2 to R3, switching is not feasible, and so he does not pursue that incentive. So pursuit of the incentive to switch from R1 to R2 does not lead beyond R2. Pursuit of the incentive to switch from R1 to R2 is not futile. The difference in the futility of incentives to switch from R1 is a difference in the grounds of choiceworthiness for R1 and R2. R1 is choiceworthy in the first decision problem but not in the second.

To obtain an acceptable version of principle alpha, we can modify its proviso. We can make it require constancy in the grounds of choiceworthiness for options common to the original and reduced decision problems. Then coherence does require that an option that is a rational choice in the original decision problem be a rational choice in the reduced decision problem.

5

Strategic Equilibrium

This chapter uses our new account of self-support to obtain a new account of equilibrium in games. It introduces strategic equilibria and shows that all of the games we treat have one.

A strategy is self-supporting if and only if given its realization there is no sufficient *reason* to switch. Hence a strategy is self-supporting only if given its realization there is no sufficient *incentive* to switch. If a strategy avoids this type of self-defeat, we call it strategically self-supporting, and a profile made up of strategies that jointly have this type of self-support we call a strategic equilibrium:

A *strategic equilibrium* is a profile of strategies that are strategically self-supporting given the profile.

Strategic equilibrium is not equilibrium with respect to all reasons. Principles of equilibrium selection, for instance, may give an agent a sufficient *reason* to switch from his part in one equilibrium to his part in a better equilibrium even if he does not have a sufficient *incentive* to switch from his part in the first equilibrium. For simplicity, however, we study only strategic equilibrium. Investigating strategic equilibrium allows us to focus on reasons that can be defined in terms of incentives.[1]

Although strategic equilibrium may ignore some reasons to switch, being a strategic equilibrium is a necessary condition for being an unqualified equilibrium. An unqualified equilibrium is a profile of strategies that are self-supporting given the profile. Thus it is a profile in which every strategy at a minimum is strategically self-supporting given the profile, that is, a profile in which no agent has a sufficient incentive to switch strategy given the profile. Being an equilibrium with respect to incentives is necessary for being an equilibrium with respect to reasons in general. Since being an unqualified equilibrium is a necessary condition for being a solution, being a strategic equilibrium is also a necessary condition for being a solution. Identifying strategic non-equilibria is thus helpful in disqualifying profiles as solutions, and therefore in identifying solutions.

---

[1] We do *not* introduce strategic solutions. Since we do not try to characterize solutions precisely, we do *not* put aside reasons that are not incentives when considering solutions.

We call the new types of self-support and equilibrium strategic because, as strategic reasoning does, they look beyond incentives to switch strategies to responses to pursuit of those incentives. Other apt, but longer, names for strategic self-support and strategic equilibrium are "self-support with respect to incentives" and "equilibrium with respect to incentives" since both restrict reasons to incentives. Sometimes for brevity we do not explicitly say that self-support and equilibria are strategic.[2]

Strategic self-support is fleshed out in terms of paths of incentives. Generation of a terminating path of incentives is a sufficient condition for a sufficient incentive. To obtain characterizations or working definitions of strategic self-support and equilibrium, we restrict ourselves to games in which it is both necessary and sufficient. Accordingly, a strategy is strategically self-supporting if and only if it does not initiate a terminating path of incentives, and a strategic equilibrium is a profile composed of strategies none of which initiates a terminating path of incentives given the profile. For games meeting the restriction, these characterizations in terms of paths are equivalent to the canonical definitions in terms of sufficient incentives.

A strategic equilibrium is a relative equilibrium, a profile of strategies strategically self-supporting relative to the profile. Relative equilibria form the foundation for the relative solutions we seek, and we focus on them. Still, we have a side interest in nonrelative strategic equilibria, that is, profiles of strategies strategically self-supporting in a nonrelative way. Being a nonrelative strategic equilibrium is a necessary condition for being a nonrelative solution, that is, a profile of nonconditionally rational strategies, and such profiles are significant from the perspective of decision theory.

The demonstration that strategic equilibria exist, in contrast with demonstrations for Nash equilibria, does not rely on mixed strategies. Strategic equilibrium does not require a probabilistic framework, although it is not inimical to one. We hold that a strategic equilibrium exists in every ideal game but demonstrate this only for normal-form games with a finite number of agents. (For brevity, we generally take it as understood that the number of agents is finite.) We cannot extend our results to other games until we provide an analysis of those games and explain how our account of strategic equilibrium applies to them, too wide-ranging a task to undertake here.

5.1. PATHS OF RELATIVE INCENTIVES

A strategic equilibrium is a profile whose strategies are self-supporting *given the profile*. What matters, then, is self-support *given a profile*, the sufficiency

[2] Selten (1988: v) uses the term "strategic equilibrium" as the subject heading for some essays on equilibrium in extensive-form games in which strategic reasoning typically arises. We appropriate the term for equilibrium in normal-form games because similar problems of strategic reasoning arise in them.

of incentives *given a profile*, and thus paths of incentives given a profile. To obtain an account of strategic equilibrium from the previous chapter's material on terminating paths of incentives, we must describe agents' incentives and paths of incentives relative to profiles. Chapter 4 did not relativize incentives to profiles, even when it considered incentives in games. To apply the principle of self-support to strategic equilibria, we must carry out this relativization.

5.1.1. Relativization to Profiles

In Chapter 4 the incentives in an agent's path were with respect to the assumption that a strategy is realized. To determine whether given a strategy an agent has an incentive to switch, we did not simply compare the strategy with the agent's other strategies. We compared it with his other strategies under the supposition that it is adopted. Similarly, to determine whether an agent has an incentive to switch strategies given the more detailed assumption that his strategy is realized as part of a particular profile, we compare his strategy in the profile with his other strategies under the supposition that the profile is realized.

The supposition that an agent's strategy in a profile is realized takes account of some of the information carried by the supposition that the profile is realized. It takes account of information that the agent's part in the profile is realized. But it leaves out information about the strategies of other agents carried by the supposition that the whole profile is realized. The difference in information carried may make an agent's incentives given the profile differ from his incentives given the strategy. Consequently, an agent's path of relative incentives away from a strategy in a profile may terminate even though his path of nonrelative incentives away from the strategy does not terminate. So a relative incentive to switch from a strategy in a profile may be a sufficient incentive even though a nonrelative incentive to switch from the strategy is not a sufficient incentive.

To illustrate, take the ideal normal-form game depicted in Figure 5.1, where arrows indicate payoff-increasing switches and so relative incentives to switch. Since incentives are relative to profiles, we represent them using profiles, not just strategies. Row has the following path of relative incentives: (R2, C2), [(R1, C2)] (R1, C1), (R2, C1). The step from (R2, C2) to (R1, C1) assumes that Column switches from C2 to C1 given (R1, C2). But since Column's switch is not part of Row's path, we put brackets around (R1, C2). Given (R1, C1) Row has an incentive to switch from R1 to R2 so that (R2, C1) results. Column has no incentive to switch strategy given (R2, C1). Since (R2, C1) has Column's response to Row's switch from R1 to R2, and since Row has no incentive to switch strategy given (R2, C1), Row's path

132

Figure 5.1   Same strategy, different profile.

terminates with (R2, C1). His path terminates although the first and last nodes assign him the same strategy, R2. The last node puts R2 in a different context than the first node.

The paths of incentives that determine equilibria have incentives that are relative to a profile. In a path of incentives for an agent, an incentive to switch from a strategy is relative to a profile containing the strategy. An agent's path indicates incentives for an agent to switch from his part in a given profile. Paths of relative incentives differ from the paths of nonrelative incentives examined in the previous chapter. The paths may depict different responses to the same strategy, and subsequently different incentives, if the strategy occurs in different profiles. A strategy may meet varying responses from other agents depending on the profile in which it is realized and, in virtue of meeting those varying responses, may generate varying incentives to switch. The response of other agents to an agent's strategy may be different for different profiles containing the strategy, and his incentives depend on their response.

Incentives given a strategy in a profile depend on the profile as well as the strategy. The incentives of an agent and the strategies of his that are self-supporting are relative to the profile assumed realized. The switches of other agents between switches of an agent in a path of incentives give the responses of other agents to the agent's strategy in the context of a profile. The responses of the other agents are relative to the profile in which the agent's strategy appears. The agent's beliefs about the responses generate his incentives to switch strategy, which in turn generate the paths of incentives with respect to which his strategies are assessed for self-support. We assume that for every strategy for every agent, other agents have a response to the strategy and that agents know about the responses to their strategies in ideal games. This knowledge is part of their knowledge of the game. Since relative responses play an important role in the generation of incentives, we take knowledge of them to be included in ideal agents' knowledge of the game.

The terminating path of relative incentives away from R2 in our example does not ensure a terminating path of nonrelative incentives away from R2. A terminating path of nonrelative incentives never has the same strategy at initial and terminal node. Whether Row has a terminating path of

133

nonrelative incentives away from R2 depends on Row's beliefs about Column's responses to his strategies. If Row believes that Column responds to R2 with C1, then given R2 Row has no incentive to switch and R2 does not start a terminating path of nonrelative incentives.[3] But if Row believes that Column responds to R2 with C2, then R2 does generate an incentive for Row to switch to R1. Given relentless pursuit of incentives, R2 starts an endless path: R2, R1, R2, ... If Row resists his incentive to switch from R1 to R2, that path terminates with R1.

An agent, for each of his strategies, predicts a nonrelative response by other agents. The nonrelative response may depend on the profile actually realized but is independent of counterfactual suppositions of a profile. The nonrelative response depends on the supposition of the strategy, but not on the supposition of a profile as context for the strategy. Nonrelative incentives obtain with respect to nonrelative responses to strategies, and not with respect to responses to strategies in the context of a profile. Given prescience the assumption of a strategy amounts to the assumption of the profile consisting of the strategy and the nonrelative response to it. But if a different profile containing the strategy is supposed, one in which the *remainder* of the profile – the part of the profile besides the strategy – is not the nonrelative response to the strategy, supposition of the profile is not equivalent to supposition of the strategy.

It may be tempting to say that an agent's path of incentives does not terminate if the path returns to the initial strategy of the path, as in the example. An agent controls only his part of a profile, so one may think he does not have a sufficient incentive to depart from a strategy if pursuit of incentives brings him back to the strategy, even if the context for the strategy is different. However, taking a path's incentives to be relative rather than nonrelative requires a different view of a path's termination. The nodes of a path of relative incentives are profiles, not strategies in isolation. They present a strategy in the context of a profile since a relative incentive obtains under the assumption that a profile is realized. In ideal games, where agents are prescient, that assumption carries information for an agent. Under the assumption, an agent knows that the profile is realized. Consequently, his incentives are with respect to that knowledge. Since the nodes of an agent's path are profiles, the same strategy may generate different incentives from one appearance in a path to the next, depending on the profile in which it appears. For one appearance of the strategy there may be an incentive to switch to another strategy, whereas for another appearance the incentive may be absent. An agent's path of incentives does not terminate if it returns to

---

[3] In a nonideal version of the game, in which Column but not Row is prescient and Row erroneously believes that R2 meets with C1, (R2, C2) is a nonrelative equilibrium.

the initial strategy in the same profile and so cycles. But if it returns to the initial strategy in a different profile, it may terminate. A switch away from a strategy that eventually returns to the strategy in a different profile is not always futile since the new context may make the strategy pay more. An agent may have a sufficient incentive to switch strategy, even if he returns to the strategy, as long as his return is in the context of a different profile.

A puzzle about our example may arise. Row's terminating path away from R2 given (R2, C2) entails that (R2, C2) is not a relative equilibrium. The profile contains a strategy that is not self-supporting given the profile. This result agrees with intuition. Intuitively, (R2, C2) is not a solution. It is no surprise that the profile is not an equilibrium. As it turns out, (R2, C1) is the unique relative equilibrium and thus the unique relative solution. But Column is indifferent between (R2, C1) and (R2, C2). Why should Column adopt C1, her part in the equilibrium profile (R2, C1)? Why shouldn't she switch to C2?

The reason is that she knows that if she switches to C2, Row adopts R1 rather than R2. Then she prefers C1. In other words, given C1 she has a terminating path of relative incentives from C2 to C1. A full specification of her path gives the context of those strategies. It runs as follows: (R2, C1), [(R2, C2)] (R1, C2), [(R1, C1)] (R2, C1). The profiles in brackets are the ones Column's switches produce before Row has a chance to respond. They are not part of Column's path but are displayed to exhibit the path's assumptions. Column's path after the initial profile gives Row's responses to her strategies. In this example Column's path stops at (R2, C1) because given the profile Column has no incentive to switch away from C1. Row's response to C1 in the context of that profile is R2, and that response generates no incentive for Column to switch strategy.

In general, there are important differences between paths of relative and nonrelative incentives. In a path of relative incentives, a strategy may appear in one profile followed by one relative response, and then appear in another profile followed by a different relative response. This cannot happen in a path of nonrelative incentives. There is at most one nonrelative response for each strategy. Also, a path of relative incentives may start with a profile P1, switch to another profile P2, and then terminate. But P2 may not contain the nonrelative response to the agent's strategy in P2. Another profile, P3 may be the nonrelative response profile for the agent's strategy in P2. Given the nonrelative response to his strategy, the agent may have a nonrelative incentive to switch strategy. His path of nonrelative incentives may not terminate. This happens for Row in the game of Figure 5.1 given relentless pursuit of incentives. His path of relative incentives away from (R2, C2) terminates in (R2, C1). But given a belief that Column responds to R2 with C2, his path of nonrelative incentives from R2 cycles endlessly: R2, R1, R2, ...

135

To illustrate the difference between relative and nonrelative equilibria, consider (R2, C1). It starts no paths in the payoff matrix, and so no paths, in particular, no terminating paths, in the completely reduced incentive structures of agents. It is a relative equilibrium. On the other hand, suppose that Row also believes that Column responds to incentives given his strategies. He believes that she responds to R1 with C1. But, in addition, he believes, as in the previous paragraph, that she responds to R2 with C2. Then if Row pursues incentives relentlessly, his path of nonrelative incentives away from R1 is R1, R2, R1, ... Row's path of nonrelative incentives does not discriminate between R1 in the context of C1, and R1 in the context of C2; it allows termination in a strategy only, and not termination in a profile such as (R2, C1). Suppose that Row resists his nonrelative incentive to switch from R1 to R2. Then he has a terminating path of nonrelative incentives to switch from R2 to R1. In this case (R2, C1) is not a nonrelative equilibrium. Row has a terminating path of nonrelative incentives away from his part in the profile. Although (R2, C1) is a relative equilibrium, it is not a nonrelative equilibrium.

Moreover, (R1, C1) is a nonrelative equilibrium in this case. Column has no nonrelative incentive away from C1, and Row only has a resisted nonrelative incentive away from R1. Neither has a path of pursued nonrelative incentives away from his part in the profile. However, (R1, C1) is not a relative equilibrium. Row has a terminating path of relative incentives away from the profile, namely, (R1, C1), (R2, C1).

Now let us formulate a general procedure for obtaining paths of relative incentives. Because we work with relative equilibria, incentives, and responses, we represent paths of incentives by paths of profiles, not strategies. In paths of relative incentives for an agent, the nodes of the path are profiles, not strategies of the agent by themselves. Given a strategy, an incentive to switch depends on the response of other agents, and their response depends on the context or profile in which the strategy appears. An agent's path of relative incentives away from a strategy takes account of the context of the strategy, namely, the profile containing the strategy. A profile in the path gives the strategy of the agent in the context of a profile.

The initial profile gives the initial assumption with respect to which incentives are taken. For nonrelative incentives a profile is inferred from the initial strategy of the path. For relative incentives the initial profile of a path is simply specified. Relative to the initial profile, an agent's incentives are with respect to the remainder of the initial profile. The agent's incentives with respect to the profile and his inferences about other agents' responses to his strategies determine subsequent strategies. Subsequent strategies in the path are assumed in the context of a whole profile that sets the stage for later incentives.

The only switches in profile that an agent can realize are switches accomplished by changing his strategy. A switch in strategies by one agent causes a change in strategy profile, so an agent's incentive to switch strategies may be represented as an incentive to switch profiles, where the agent's strategy changes from the first profile to the second but other agents' strategies are constant. An agent's incentive to switch profiles is a preference for the second profile given the first in cases in which the two profiles are alike except for the agent's strategy. Instead of putting the two profiles that represent an agent's incentive in his path of incentives, we put in the first profile and the profile that gives the response of other agents to his switch in the context of the profile his switch produces. Thus the second profile of an agent's path is not necessarily the *switch profile* obtained by shifting from the agent's initial strategy to another strategy while the strategies of other agents are held constant. Rather the second profile comprises the agent's new strategy and the relative response of other agents to that strategy. The agent's new strategy and the relative response to it constitute the *response profile* for the strategy switch relative to the original profile.

A response to a strategy is the reaction of other agents to incentives that arise given the strategy in the context of a profile. A profile in an agent's path after the first profile has a strategy of the agent and other agents' response to it. An agent's path progresses from a profile containing a strategy of his to a profile introducing a new strategy for him and other agents' response to that strategy and to their own strategies. For an agent, initial incentives are with respect to the initial profile, and subsequent incentives are with respect to response profiles. After the first strategy, the agent's strategy appears in a profile along with the response to it.

For each agent, a profile has a strategy for the agent and strategies for the other agents. The response to the agent's strategy given the profile is the remainder of the profile if the other agents lack incentives to switch strategies given the profile. If given the profile the other agents have incentives to switch strategies and they pursue them, then the response to the agent's strategy is the result of their pursuing those incentives. Incentives with respect to a profile containing a strategy may therefore differ from incentives with respect to the response profile for the strategy in the context of the original profile. The relative response to a strategy in a profile is not necessarily the same as the remainder of the profile. Given a strategy in a profile the remainder of the profile may differ from the response to the strategy in the context of the profile. There is a difference between the other agents' strategies accompanying an agent's strategy in a profile and the other agents' response to the agent's strategy relative to that profile. The other agents' response is the subprofile they realize after they finish pursuing incentives given the agent's strategy in the context of the original profile. A profile contains the other agents' response to his strategy only after their strategies cease changing.

To obtain a path of relative incentives for an agent in an ideal game, where incentives follow payoff increases, construct a series of profiles as follows: First, take an arbitrary profile; second, a profile that differs from the first only in the strategy of the agent and such that a switch from the agent's strategy in the first profile to his strategy in the second increases his payoff given that the strategies of other agents are constant; third, a profile in which the agent's strategy is the same as in the second profile, and the strategies of other agents change one by one in payoff-increasing ways; fourth, a profile in which only the agent's strategy changes, and it changes so that the agent's payoff increases given that the strategies of other agents are constant; and so on, with payoff-increasing changes in strategy for the agent alternating with the net effect of individual payoff-increasing changes in strategy for other agents. After such a series of profiles is obtained, compress it by retaining only the initial strategy in the initial profile and the response profiles for the agent's subsequent strategies. After the compression, no two adjacent profiles assign the same strategy to the agent. The compressed series of profiles omits switch profiles rather than response profiles because the switch profiles can be restored easily. To obtain the switch profile between two profiles in an agent's path of incentives, just put the agent's second strategy into the first profile.

Bear in mind that sometimes a switch profile is also a response profile. This happens when there is no reaction to the switch profile by other agents. Then the switch profile is retained. Also note that the foregoing method of constructing paths of incentives puts aside cases where the initial sequence of incentives has two adjacent switches by the same agent. Rationality requirements may prohibit such switches since there is no change in the agent's grounds of preference if the strategies of other agents are constant. The issue is open; see Section 4.4. In examples we rule out such switches for simplicity, unless entertaining criticisms as in the case of Figure 8.9.

A path of relative incentives that terminates in a strategy terminates in a strategy in a context. That is, it terminates in a profile, not the strategy taken in isolation. Likewise a path of relative incentives that terminates in a tree terminates in a tree of profiles, where incentives of the tree are relative to profiles, and the tree is closed with respect to incentives relative to profiles. Fortunately, equilibria in a game depend on the completely reduced relative incentive structures of agents. These structures have some convenient features. In them all trees are paths, and all paths terminate in a profile. Moreover, a path of relative incentives for an agent terminates if and only if it is a closed path that has a last strategy for the agent. The terminal profile contains the agent's last strategy and the response of other agents to

138

$$1, 2 \qquad 2, 1$$

$$2, 1 \qquad 2, 1$$

$$2, 1 \qquad 1, 2$$

Figure 5.2   Paths of relative incentives.

it. The agent has no incentive to switch from his last strategy given the other agents' response.

5.1.2.   Features of Relativized Paths

To illustrate paths of relative incentives further, consider an ideal normal-form game that instantiates the payoff matrix in Figure 5.2. Since the game is ideal, incentives follow payoff increases. Given (R1, C2), Column's payoff increases if she switches from C2 to C1, in other words, if she switches from (R1, C2) to (R1, C1). Suppose Row's response to (R1, C1) is R2, that is, a move to (R2, C1). Given Row's response, Column cannot gain by switching strategies again. So Column's path from (R1, C2) to (R2, C1) terminates with (R2, C1). But if Row responds to (R1, C1) with R3 instead of R2, and to (R3, C2) with R1, Column has the following path of incentives: (R1, C2), [(R1, C1)] (R3, C1), [(R3, C2)] (R1, C2), ... The profiles in brackets are the switch profiles omitted during compression of the initial sequence of profiles.

The two payoff-increasing switches available to Row given (R1, C1) generate different paths of incentives for Column, namely, (R1, C2), [(R1, C1)] (R2, C1) and (R1, C2), [(R1, C1)] (R3, C1), [(R3, C2)] (R1, C2), ... Suppressing the switch profiles, the two paths are (R1, C2), (R2, C1) and (R1, C2), (R3, C1), (R1, C2), ... Finally, suppressing Row's responses, we obtain two paths of strategies for Column: C2, C1 and C2, C1, C2, ... The two paths begin with the same strategies for Column but then differ as a result of differences in responses to those strategies. The first path for Column terminates with C1, but the second does not.

Each path of incentives for Column is part of a different concrete realization of the payoff matrix. Column's path of incentives away from (R1, C2) depends on the path of payoff increases that gives Row's actual responses. We do not say that there is one path of incentives for Column from C2 to C1 and that it does not terminate with C1 because there is a way in which it may return to C2. We say that there are two *potential* paths of incentives for Column from C2 to C1, and one of them terminates with C1 whereas the other returns to C2. Once a concrete realization of the payoff matrix is fixed, only one of these potential paths is actual. A normal-form game is a concrete realization of a payoff matrix, and the concrete realization settles the responses to an agent's strategies that determine the agent's paths of incentives in the game. The agent's paths of relative incentives depend on other agents' responses to the agent's strategies. In an ideal game the response to an agent's strategy given a profile follows payoff increases, but not every series of payoff increases for other agents yields their response and so a path of incentives for the agent. At most one profile gives the relative response to an agent's strategy.

An agent's paths of relative incentives are dependent on other agents' responses to his strategies. Relative incentives for an agent are with respect to the agent's knowledge of his and other agents' pursuit of incentives relative to profiles, full knowledge in ideal games. In ideal games every path of incentives for an agent follows payoff increases for the agent and involves payoff-increasing responses by other agents. Since all incentives to switch follow payoff increases, an agent's path of incentives stands for a path of payoff increases for agents through the payoff matrix. But agents do not pursue every payoff increase open to them. A possible path of incentives for an agent that involves unpursued payoff increases for other agents is not a genuine path of incentives for the agent. The agent is aware of the payoff increases that other agents pursue and knows that the possible path does not incorporate actual responses by other agents. Only pursued payoff increases yield responses and hence the incentives for an agent that hold in light of other agents' responses to his strategies.

The payoff matrix and the set of possible patterns of agents' pursuit of payoff increases through the matrix are independent of assumptions about responses. Consequently, the set of possible paths of incentives for an agent is independent of responses. A single profile may initiate many possible paths of incentives for an agent. Each possible path corresponds to a different possible pattern of pursuit of incentives by the agents. The possible paths differ according to the incentives the agent pursues and the various possible responses of other agents to his strategies. The possible paths are distinguished by the profiles that form their nodes. The profiles give the agent's strategy and other agents' possible responses. Different possible paths of incentives may involve the same strategies but advance different profiles as the context for those strategies.

Since incentives are relative to profiles and some paths have many profiles, one wonders whether the profiles prior to a node in a path have any influence on the incentives at the node. In an ideal game an agent has foreknowledge of the profile realized, whatever it is, and that foreknowledge has priority over other idealizations with which it may conflict. So the epistemic consequences of supposing a profile are the same whatever the context for

140

the profile. As a result, the response to an agent's strategy relative to a profile is the same no matter what the context for the profile, and so no matter where it appears in the agent's path.

To see this point in more detail, consider the following: In ideal games the agents are ideal and conditions are ideal for strategic reasoning; in particular, all agents are rational, prescient, and are fully informed about the game. Since agents are prescient and informed, incentives in an ideal game follow payoff increases. Incentives in ideal games track payoff increases, and payoff increases with respect to a profile depend only on the profile. Relative incentives are just payoff increases with respect to the profile supposed. Given a hypothetical profile, beliefs and preferences are the same whether or not supposition of the profile is embedded in supposition of another profile. The supposition of a profile does not alter the payoff structure, so in an ideal game it does not alter the incentive structure either. The relative incentives of agents are constant whatever profile is supposed. The supposition that a profile is realized does not alter the incentive structure even if that profile is not the one realized. No change in profile supposed affects the incentive structure. The change affects only the profile with respect to which relative incentives are computed.

Thus in ideal games incentives depend only on the profile supposed and not prior suppositions of profiles. In ideal games supposition of noninitial profiles of an agent's path of relative incentives swamps out the effect of the initial profile's supposition. The initial profile merely serves as a starting point for the path. It does not influence incentives with respect to subsequent profiles. The only relevant context for an incentive to switch from a strategy given a profile is the profile itself. Relative incentives to switch strategy are independent of everything except the profile supposed. An incentive relative to a profile depends only on the profile, and not on suppositions of other profiles earlier in the path of incentives.

This feature of ideal games has important consequences concerning the justification of solutions. Thanks to the independence of the epistemic consequences of supposition of a profile, reasoning involving supposition of profiles can justify the profile actually realized. Since the epistemic consequences of a hypothetical profile do not vary according to the profile actually realized, one can infer the profile to be realized from the epistemic consequences of hypothetical profiles. One does not have to use the profile actually realized – which is assumed to be justified in ideal games – to infer the epistemic consequences of hypothetical profiles. Circularity can be avoided.

Now let us explore a final point about paths of relative incentives. There are two ways to conceive of paths of relative incentives away from a profile: According to one view, there is no path for an agent unless given the profile the agent has an incentive to switch strategies. According to the other view,

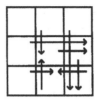

Figure 5.3   Two types of path.

there is a path for an agent, even if given the profile the agent has no incentive to switch strategies, provided that other agents respond to the agent's strategy given the profile, that is, pursue their incentives given the profile, and considering their response the agent has an incentive to switch strategies. The first interpretation computes the agent's incentives with respect to the profile given, whereas the second interpretation computes the agent's incentives with respect to the response profile, that is, the profile obtained after substituting the other agents' response to his strategy for their strategies in the profile given.

To illustrate the difference between the two ways of obtaining paths of incentives, consider (R2, C1) in an ideal game with the incentive structure depicted by Figure 5.3. (This incentive structure is generated by the game with an unattractive Nash equilibrium in Figure 3.3 and reappears in Figure 7.4.) (R2, C1) is not the response profile for Row's strategy given the profile. (R2, C3) is the response profile. It gives Column's response to R2 given (R2, C1). So if Row's incentives are taken with respect to response profiles, then Row has a path away from (R2, C1) because he has an incentive to switch from (R2, C3) to (R3, C3), even though he has no incentive to switch from (R2, C1) to another profile containing C1.

The way the interpretation of paths is settled is significant. In some games it determines whether a profile counts as an equilibrium. To see this, imagine a profile with the following features: Under the first interpretation, taking the profile as is, no agent has a terminating path away from the profile. One agent has no path away from the profile, and another agent has a nonterminating path away. But under the second interpretation, using incentives with respect to the response profile, the first agent has a terminating path away from the profile since the response of the second agent to the profile moves the first agent to another profile from which the first agent has an incentive to switch. This incentive starts a terminating path for the first agent. In this case the profile is an equilibrium under the first interpretation, but not under the second interpretation.

In evaluating a proposed nonrelative equilibrium or a relative equilibrium in a nonideal game, the incentives of agents may appropriately be taken with

142

respect to different profiles. But in evaluating a proposed relative equilibrium in an ideal game, initial relative incentives for different agents should all be with respect to the same profile. Under the first interpretation of relative incentives, the initial profile serves as a benchmark for all initial relative incentives. Under the second interpretation, no profile serves as a benchmark for initial relative incentives of all agents. We adopt the first interpretation to obtain a benchmark appropriate for initial relative incentives in ideal games. Initial relative incentives to switch away from a strategy given a profile containing the strategy are taken with respect to the profile itself, not with respect to the response profile. We attend to an agent's incentives with respect to the initial profile in a path, and not his incentives after substituting the other agents' response to his strategy in the initial profile for the remainder of the initial profile. The initial profile minus the agent's strategy gives the context for his strategy. Initial incentives to switch are determined with respect to it.

Initial and subsequent incentives in an agent's path of relative incentives have different roles. An initial profile is taken as a proposed solution to the game. An initial strategy in the context of the profile is examined as part of the profile's evaluation as a solution. What matters for evaluation of a profile as a solution are an agent's incentives given the profile, and not given the profile obtained by replacing the remainder of the profile with the other agents' response to his strategy. On the other hand, a subsequent strategy is not proposed in the context of a profile as part of a solution. A subsequent profile is not proposed as a solution, but as a means of pursuing hypothetically the result of a switch in strategy. It has a role in strategic reasoning. It is used to determine the ultimate result of an agent's rejecting the proposed solution by switching strategies. Subsequent profiles present responses to the agent's strategies and reach the profile ultimately realized if the agent switches away from his strategy in the initial profile. A terminating path away from his initial strategy explains why the path's terminal profile is ultimately realized if he switches. Subsequent incentives arise in virtue of the other agents' responses to his switches in strategy. Pursuing a relative incentive involves moving from one profile to another by substituting a new strategy in the initial profile. The agent's next incentive depends on other agents' responses to his new strategy. The response profile for the new strategy is in the agent's path of relative incentives, and his next incentive, if any, is with respect to it. After a switch by the agent, the remainder of the profile is replaced with the response to his new strategy to obtain the context for subsequent incentives to switch strategy. An agent's relative incentives after a switch in strategy are appropriately taken with respect to the response profile, the profile containing his new strategy together with the other agents' response to it. This interpretation of subsequent incentives provides for their role in strategic reasoning.

143

Thus the context for incentives relative to a profile depends on whether the profile is the initial profile of a path or a subsequent profile. If it is the initial profile, the context is the remainder of the profile. This yields incentives with respect to the profile. If it is a subsequent profile, the context is the remainder of the response profile. This yields incentives with respect to the response profile. Consequently, the terminal strategy of a terminating path generates no incentives to switch with respect to the response profile for the terminal strategy. The terminal profile has the other agents' response to the agent's terminal strategy. The agent has no incentive to switch from his terminal strategy given the other agents' response to it. He has no incentive to switch given the response profile, but not necessarily given the switch profile.

Now that paths of relative incentives have been introduced, we apply Chapter 4's distinctions concerning incentive structures. An agent's paths of *pursued* relative incentives form his *completely reduced* structure of relative incentives. This structure is explained by the agent's *unreduced* and *partially reduced* structures of relative incentives. The former has all his relative incentives, and the latter has the relative incentives he pursues given relentlessness. Analogues of the selection and stopping rules govern the transitions from unreduced to partially reduced structures, and from partially reduced to completely reduced structures. Evaluation of a profile as a strategic equilibrium appeals to agents' completely reduced structures of relative incentives.

5.2. EXISTENCE OF EQUILIBRIA

From the definition of strategic equilibrium, we obtain the following equivalent characterization for games where nongeneration of terminating paths of incentives is both necessary and sufficient for strategic self-support given a profile:

A *strategic equilibrium* is a profile of strategies that do not start any terminating paths of incentives given the profile.

The rest of this chapter uses this characterization to argue that a strategic equilibrium (not necessarily unique) exists in every ideal normal-form game with a finite number of agents. It is important to establish the existence of strategic equilibria in these games since a major reason for rejecting Nash equilibrium in favor of strategic equilibrium is the nonexistence of Nash equilibria, or profiles of jointly incentive-proof strategies, in some ideal games with a finite number of agents. Strategic equilibrium has the same shortcoming unless strategic equilibria exist in these games. Before the next section presents the existence proof, this section addresses some preliminary

144

matters. It examines some failed attempts to demonstrate the existence of equilibria and explains the role of our assumption that games are ideal.

5.2.1. *Failed Arguments*

Before presenting the argument for the existence of equilibria in ideal normal-form games with a finite number of agents, let me present some arguments that fail. They elucidate the nature of joint self-support and reveal impediments to establishment of its possibility.

The outcome of an ideal game depends on the principles of rationality the agents follow and their circumstances, in particular, their beliefs and desires. We cannot use the principles of rationality and the circumstances of the agents to determine the outcome of an ideal game, and then show that it is an equilibrium. Determining the outcome goes beyond this book's scope. However, without determining the outcome, one may argue as follows that it must be an equilibrium, that is, a profile of jointly self-supporting strategies:

A profile has jointly self-supporting strategies if and only if the strategies are self-supporting given the profile. Prescience entails foreknowledge of the profile realized, and foreknowledge turns joint self-support into self-support. For the profile realized, self-support given the profile is the same as self-support. Since ideal agents follow the principle of self-support, the outcome has to be jointly self-supporting, and therefore an equilibrium.[4]

Unfortunately, this simple argument has a flaw. It begs the question. It assumes that an equilibrium exists. For ideal agents cannot be counted on to comply with the principle of self-support if joint self-support is impossible. To assume they achieve self-support is to assume that an equilibrium exists. We can assume that agents achieve rationality and self-support if possible. But we cannot assume that they achieve rationality and self-support, period, without begging the question about an equilibrium's existence. The argument can show only that an equilibrium is realized if possible.

The next argument is similar but has a slightly different approach to self-support. It uses Chapter 4's result that in every decision problem an agent has a strategy that is self-supporting:

In all games, ideal and nonideal alike, every agent has a self-supporting strategy. Taken together these strategies form a profile of self-supporting strategies. Such a profile is an equilibrium. Hence every game has an equilibrium.

The overstrong conclusion is a sign that the argument is flawed. The argument shows the existence of only a *nonrelative* equilibrium, that is, a profile

---

[4] This argument is inspired by a theorem, proved by Aumann and Brandenburger (1991: 4), about the realization of a Nash equilibrium in games in which agents are rational and have foreknowledge of the profile realized. The argument's flaws in no way discredit the theorem.

of self-supporting strategies. We want to show the existence of a relative equilibrium, and a profile of self-supporting strategies is not necessarily a *relative* equilibrium. A relative equilibrium is a profile of strategies that are self-supporting *given* that the profile is realized. Given a profile's realization, an agent's information may change and alter his incentives and the strategies that are self-supporting. This happens in the nonideal game of Figure 7.6, which has profiles of self-supporting strategies, but no profile of strategies that are self-supporting given the profile. If a profile of self-supporting strategies is realized, some strategies lose self-support. Although a nonrelative equilibrium exists, no relative equilibrium exists.

One may try to amend the argument this way:

In an ideal game all agents have foreknowledge of the profile realized. So supposition of the profile realized does not change information, and so does not change incentives or self-support. Moreover, the profile realized is a profile of self-supporting strategies. So it is also a profile of strategies that are self-supporting given the profile. In other words, the nonrelative equilibrium realized is also a relative equilibrium.

The flaw here is the assumption that the profile realized is made up of self-supporting strategies, or the assumption that a nonrelative equilibrium is realized. The ideality of the agents does not warrant the assumption, since their ideality guarantees only that they adopt jointly self-supporting strategies *if possible*. We have shown that a nonrelative equilibrium exists, but not that it is realized, or that if it is realized the strategies in it are still self-supporting. The assumption that the profile realized is made up of self-supporting strategies again begs the question. It presumes that there is a profile such that if it is realized each strategy is self-supporting. But this is what we are trying to show.

Let us look at another flawed attempt to use the existence of a profile of self-supporting strategies to show the existence of an equilibrium. To justify the assumption that some profile of self-supporting strategies is realized, the argument takes advantage of the fact that the games we treat are normal-form games. It tries to use the causal independence of agents' strategies to overcome the problem of the relativity of self-support:

An equilibrium is a profile of strategies that are jointly self-supporting. The only way that a profile of self-supporting strategies can fail to be jointly self-supporting is if the information agents have differs from the information they have if the profile is realized. The realization of strategies by other agents may give an agent new information in some games, such as multistage games where the agent observes the moves of other agents before he moves. But in normal-form games the strategies realized by other agents do not give an agent new information. He observes the strategies of other agents only after he and the other agents adopt strategies and the game has been completed. As a result, there is no difference between the information with respect to which an agent's strategy is self-supporting taken by itself and the information with respect to which it is evaluated for self-support when taken together with self-supporting strategies of other agents. His strategy's self-support is independent of other agents' strategies. Their strategies do not provide him with new information, and so do not change incentives to switch. Strategies that are self-supporting taken individually are self-supporting taken together. So profiles of self-supporting strategies have jointly self-supporting strategies. Because every normal-form game has a profile of self-supporting strategies, it also has an equilibrium.

This argument's crucial assumption is that in normal-form games the agents' information is independent of the profile supposed. The assumption is false for normal-form games where agents are prescient. Prescience entails the accuracy of predictions, even given counterfactual profiles. In normal-form games without prescience a profile may be realized without all agents having foreknowledge of its realization. The realization of a profile carries direct information to an agent about his part in the profile but no direct information about other agents' strategies. And resources for obtaining indirect information about their strategies may be insufficient. However, in games with prescience the realization of a profile carries information that it is realized to all agents. The agents' information is dependent on the profile supposed.

Although the idealization of prescience removes obstacles to the existence of equilibria by making agents' choices informed about other agents' strategies, it also creates difficulties for the preceding argument for the existence of equilibria. A strategy's being self-supporting for an agent depends on the agent's information about other agents' strategies. His information determines whether he has an incentive to switch from a strategy. In some cases for some agent a certain strategy is self-supporting, but its being self-supporting depends on his information that other agents do not adopt self-supporting strategies. In a profile of self-supporting strategies, the strategies of other agents are contrary to his expectations. If the agent is prescient, realization of those strategies carries information that alters his incentives and self-supporting strategies. Putting the agent's self-supporting strategy together with self-supporting strategies for other agents does not yield a profile such that each agent's strategy is self-supporting given knowledge of the other strategies. So under the idealization of prescience, putting together self-supporting strategies does not yield a profile of strategies self-supporting given the profile. That is, it does not yield an equilibrium.

The argument's mistaken assumption confuses a change in information with a change in information under a supposition. That is, it confuses a change within the actual world with a change from the actual world to a counterfactual world. Only one profile is actually realized and predicted; the rest are supposed realized and predicted. Their supposition carries information about other agents' strategies. But the information they carry is hypothetical, not actual, since it obtains under a counterfactual supposition. A change in an agent's information differs from a supposed change in his information as different profiles are assumed to be realized. An assumption that a profile is realized does not *change* information about other agents' strategies, but the assumption *carries* information about their strategies. In a normal-form game an agent's information is not affected by the strategies of other agents. But an agent's information under supposition of a profile may vary with the profile supposed. The supposition of the profile may cause a change in the information supposed, even if the realization of the profile would not cause a change in information. Although information is fixed, suppositions about information are not fixed. Suppositions about the agents' information change with suppositions about the profile realized. The agents' information is dependent in this way on the profile supposed.

The preceding argument shows the existence of an equilibrium only in a nonideal normal-form game in which supposition of a profile carries no information to an agent except information about his strategy in the profile, that is, no information except that carried by supposition of his strategy in the profile. In such a game the distinction between relative and nonrelative equilibrium collapses. A profile of strategies that are self-supporting is also a profile of strategies that are self-supporting given the realization of the profile. We need an argument that shows the existence of relative equilibria in ideal games, in which relative and nonrelative equilibria diverge.

The failures of the preceding arguments teach an important lesson. Strategies that are self-supporting taken separately may not be self-supporting taken jointly. Their joint supposition may carry information that changes incentives and self-support. We know from the previous chapter that each agent has a self-supporting strategy. But we do not know that a profile of self-supporting strategies is an equilibrium, that is, that each strategy is self-supporting *given the profile*. Given prescience in some cases the supposition that other agents perform certain self-supporting strategies carries information that changes the strategies that are self-supporting for an agent. To illustrate, take a two-person normal-form game and suppose that one agent believes that his opponent does not adopt a certain self-supporting strategy. The agent has a belief that his opponent adopts a certain self-defeating strategy. The strategies that are self-supporting for the agent are based on this belief. Under the supposition that his opponent does adopt a self-supporting strategy, the agent knows it given prescience. So the supposition carries new information, and the strategies of his that are self-supporting may change. Figure 7.6 provides an example in which this happens in a nonideal game where agents are prescient but one agent is irrational.

Figure 5.4   Multiple Nash equilibria.

The phenomenon may also occur in ideal games in which all agents are jointly rational if possible. Since the supposition that a profile is realized carries inform ation about its realization, the realization of a profile of strategies that are self-supporting taken individually may carry information with respect to which some of the strategies are not self-supporting. In other words, self-supporting strategies may not be jointly self-supporting. Perhaps some agent's strategy is self-supporting because he has a belief that other agents adopt self-supporting strategies different from the ones in the profile. The assumption about the agents' rationality does not rule out this possibility. Then given the joint realization of the self-supporting strategies in the profile, his belief about other agents changes, and his strategy may no longer be self-supporting.

This may happen in games with multiple Nash equilibria, for instance, an ideal normal-form game with two Nash equilibria represented by Figure 5.4. In this game a profile of self-supporting strategies from different Nash equilibria does not form an equilibrium. Although each strategy in the profile is self-supporting, the strategy is not self-supporting given the other strategy. It is not self-supporting given the profile. (R1, C1) and (R2, C2) are Nash equilibria. Suppose that for each agent and strategy, given that the agent adopts the strategy, he predicts that the other agent does what is required to achieve one of these Nash equilibria. Then R1 and R2, and C1 and C2, are self-supporting strategies. However, (R1, C2) is not a strategic equilibrium. Although each component is self-supporting taken by itself, the combination is not an equilibrium. Given (R1, C2), Column has a sufficient incentive to switch to C1. For Column, C2 is evidence of R2. But R1 is part of the assumption of (R1, C2). So given that profile, C2 indicates Column's failure to respond to a sufficient incentive to switch. Given (R2, C2), C2 is self-supporting, but given (R1, C2), C2 is not self-supporting. When (R2, C2) is assumed, the situation is ideal – all agents respond to sufficient incentives to switch. When (R1, C2) is assumed, Column is taken to make a mistake given her knowledge of the profile – she fails to respond to a sufficient incentive to switch.

Column's mistake given (R1, C2) does not contravene her ideality. Her mistake is hypothetical, not actual. Her mistake occurs under a counterfactual supposition, not in fact. By stipulation, agents are jointly rational if

149

possible. Since joint rationality is possible in this game, agents are jointly rational, and hence Column in particular is rational. Her irrationality under a counterfactual supposition does not undermine her rationality – nor her conditional rationality. Her conditional rationality conflicts with prescience, and so gives way under supposition of (R1, C2), but is robust and resurfaces given other suppositions. Column's mistake in (R1, C2) has no implications for Column's choices in other contexts. (R1, C2) does not carry information that Column is irrational into further speculations. If, for instance, Column switches from C2 to C1, we do not suppose that Row's response rests on a belief that Column is irrational.

Since the game in our example is ideal, we do not suppose that (R1, C2) is realized. We suppose that some other profile is realized. Then in the context of that supposition, we make another, embedded supposition that (R1, C2) is realized. The supposition that Column makes an irrational choice is embedded in the supposition that Column is rational. We speculate: If Column is rational, then if Column does such and such and is irrational, then . . . . Such forms of embedded supposition are not inconsistent. Embedded suppositions are not conjoined but hop from one possible world to another. Although the agents are ideal and do not make mistakes, we may entertain counterfactual situations in which they do make mistakes. That is what happens when we entertain (R1, C2). In the ideal game (R1, C2) is not realized. But the profile may be entertained counterfactually. Under its supposition, Column's choice of C2 is irrational.

5.2.2. *Existential Doubts*

The foregoing points about joint self-support not only dispose of some arguments for the existence of equilibria, but also resurrect skepticism about the existence of equilibria in ideal games. Take an ideal version of Matching Pennies. There is no Nash equilibrium because for every profile, some agent has an incentive to switch. Is there any ideal game in which for every profile some agent has a *sufficient* incentive to switch given the profile? In this case no profile, if realized, is composed of self-supporting strategies; the game fails to have a strategic equilibrium. We then face the same problem that led us to abandon the standard of Nash equilibrium. Can such cases arise?

Let us explore the issue by considering how, *putting aside ideal conditions*, games without strategic equilibria may arise. Our example is a nonideal version of Matching Pennies in which, for every profile, some agent has a sufficient incentive to switch. Later we see why this situation cannot occur in an ideal version of the game. The example brings out the importance of our idealizations.

$$1, -1 \quad -1, 1$$

$$-1, 1 \quad 1, -1$$

Figure 5.5   Matching Pennies.

Figure 5.6   Sufficient incentives to switch.

Figure 5.5 provides the payoff matrix for Matching Pennies. Consider a concrete realization of this matrix that has the following features: If (R1, C1) is realized, both agents have foreknowledge of its realization. As a result, Column has an incentive to switch to C2 given (R1, C1). She also believes that in the context resulting from her switch Row's response to C2 is R1. That is, she believes that Row does not switch in response to her switch. Her incentive to switch to C2 is therefore sufficient. She has a terminating path of relative incentives from C1 to C2. Next, if (R1, C2) is realized, both agents have foreknowledge of its realization. Row therefore has an incentive to switch to R2 given (R1, C2). He believes that in the context resulting from his switch Column's response to R2 is C2. So his incentive to switch to R2 is sufficient. He has a terminating path of relative incentives from R1 to R2. Similarly, if either (R2, C2) or (R2, C1) is realized, the agents have foreknowledge of the profile's realization, and the agent with an incentive to switch believes that the other agent does not change strategy in response to his switch so that he has a sufficient incentive to switch.

Figure 5.6 is a diagram of the agents' sufficient incentives to switch. It is just like the diagram for incentives to switch in the ideal version of Matching Pennies. For every profile, some agent has a sufficient incentive to switch. In each profile the agent who has an incentive to switch given the profile has a terminating path of incentives. His path terminates because he believes that the other agent does not respond to his switch. He believes that the other agent does not pursue incentives to switch. Because his path terminates, his incentive to switch is sufficient. Whatever profile is realized, there is a terminating path away from one strategy in the profile, and so a sufficient incentive to switch away from the strategy. This is what we fear: no equilibrium.

This example would be a counterexample to our claim about the existence of equilibria if the game were ideal. But it is not. Depending on how the example is filled out, there is a violation of either the idealization of prescience or the idealization that agents are jointly rational if possible, assuming satisfaction of the idealization of informedness.

Imagine the agents are prescient. Then suppose, for example, that (R1, C1) is realized. Column believes that if she switches to C2, Row's response is R1. If Row is rational, however, his response to her switch is R2 since given (R1, C2) Row has a sufficient incentive to switch to R2; he believes Column's strategy is C2 if he switches. But if his response to C2 is R2, then Column has a false belief about his response, and so is not prescient. Hence, given Column's prescience, Row does not pursue his sufficient incentive to switch from R1 to R2 given (R1, C2). He fails to be rational although it is possible for agents to be jointly rational.

Given prescience it is possible for both agents to pursue sufficient incentives to switch. If one agent does, then given prescience the incentives to switch of the other agent are no longer sufficient. Given that one agent believes that his opponent pursues sufficient incentives to switch, the agent's own incentives to switch are not sufficient. His paths of incentives do not terminate. He pursues sufficient incentives by default. Joint rationality requires only that one agent take the initiative to pursue sufficient incentives. It is within reach. Given prescience the example therefore presumes that neither agent pursues all sufficient incentives despite the possibility that they pursue sufficient incentives jointly. Hence given prescience the example violates the idealization that agents are jointly rational if possible.

This point assumes that each agent has accurate expectations about his opponent's responses. It assumes that prescience includes predictive power concerning hypothetical profiles, not just concerning the actual profile. Moreover, it assumes that prescience entails correct beliefs *about* hypothetical situations, not just correct beliefs *in* hypothetical situations. No agent is surprised by the other agent's response in any hypothetical situation. Each agent knows the other agent's response to each strategy in every context.

On the other hand, take a version of the example in which the agents are jointly rational if possible. In this case agents are jointly rational if possible by default. Joint rationality is not possible. Whatever profile is realized some agent fails to pursue a sufficient incentive to switch. Each agent believes the other agent fails to pursue an incentive to counteract a switch, and one agent is mistaken. Consequently, some agent is not prescient about responses to switches in strategy. Prescient agents know the responses to their strategies even in hypothetical situations.

This version of the example can be further fleshed out as follows: Take (R1, C1). Row has foreknowledge of the profile realized. So given the profile Row lacks an incentive to switch strategy. In contrast, Column has a sufficient incentive to switch given the profile. Row believes that Column

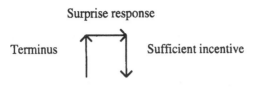

Figure 5.7 Unexpected responses.

does not pursue her incentive to switch. But if Column pursues her incentive, contrary to Row's expectations, then Row has a sufficient incentive to depart from the resulting profile, (R1, C2). Row has a path of incentives away from R1 that terminates in R2. He has no incentive to switch from R2 if, as he expects, Column then fails to pursue her incentive to switch to C1. So given (R1, C1), if Column pursues her sufficient incentive to switch to C2, contrary to Row's expectations, then Row's strategy and Column's unexpected response produce a profile given the realization of which Row has a sufficient incentive to switch strategies. Row's situation is depicted by the top and right arrows in Figure 5.7.

Now suppose, for example, that (R2, C1) is realized. Row has fore-knowledge of the profile and forgoes a sufficient incentive to switch, the one indicated by the arrow on the left of Figure 5.7. His incentive is sufficient because he thinks that if (R1, C1) is realized, Column forgoes her incentive to switch. But in fact she does not. Column does not forgo any incentives. Row is the one who forgoes incentives. Row's incentives given Column's surprise response are laid out in the preceding paragraph. Since Row is wrong about Column's response given (R1, C1), he is not prescient. He is not prescient even though he has foreknowledge of the profile realized. He lacks foreknowledge of Column's responses in hypothetical situations.

The foregoing shows that in ideal conditions the equilibriumless version of Matching Pennies does not arise. But we have to show in general that given the idealizations joint pursuit of sufficient incentives is always possible. Demonstrating this would be straightforward if we could assume that each agent in an ideal game, in calculating the other agents' response to his strategy, takes other agents to follow sufficient incentives to switch. Then a strategy self-supporting given their response would be a profile of strategies self-supporting given the profile. But the assumption that each agent counts on other agents to pursue sufficient incentives presupposes that other agents can jointly follow sufficient incentives, and this needs to be shown. We cannot assume that because the game is ideal, agents follow sufficient incentives. They do not if it is impossible for them to do this. They fail if there is no equilibrium.

This section presents a proof that every ideal normal-form game with a finite number of agents has an equilibrium. The proof avoids the problems concerning unexpected responses that undermine the foregoing attempts to demonstrate the existence of equilibria. It shows how to construct an equilibrium by putting together certain *relatively* self-supporting strategies: strategies self-supporting given a profile. It shows that because of the idealizations, certain profiles of relatively self-supporting strategies are profiles of strategies that are self-supporting given the profile, and so are equilibria. The demonstration invokes only certain specially constructed profiles of relatively self-supporting strategies.

The proof relies on the assumption that ideal agents are jointly rational if possible. Since rationality requires adopting self-supporting strategies, we assume that ideal agents adopt strategies that are jointly self-supporting if possible. In other words, they realize an equilibrium if possible. This holds for subgroups of agents, as well as the whole group of agents. Sections 2.2 and 3.1 motivate these assumptions about agents. The proof also relies on the causal independence of agents' choices in normal-form games. In these games no agent's choice has a causal influence on other agents' choices.

The proof is intended to show the existence of equilibria in concrete games. Hence the incentive structures it addresses are the completely reduced incentive structures of agents. In these incentive structures all trees are paths and all paths terminate in a strategy. But the proof itself assumes nothing about the incentive structures it treats, except that they are for the agents in an ideal game. Hence, in effect, the proof also shows the existence of an equilibrium with respect to the unreduced and partially reduced incentive structures of agents. The proof of the existence of an equilibrium is effectively independent of the type of incentive structure. To obtain an existence proof for a type of incentive structure, we merely specify application of the proof to that type of incentive structure. We present the proof with greater generality than is necessary because the existence of equilibria with respect to agents' unreduced and partially reduced incentive structures may be useful for explanations of equilibria with respect to their completely reduced incentive structures.

For the existence proof, I use a lemma concerning the terminus of an agent's path of incentives. The terminus is a tree of strategies, in special cases a single strategy. Call a member of the terminus a *terminal strategy*. The lemma says that a terminal strategy does not initiate a terminating path of incentives. This is readily established. Suppose that some strategy S is a member of a tree T that terminates a path. Because T terminates a path, it is closed, and every member of T appears in every closed subtree it immediately precedes. Suppose then that S initiates a path that terminates in a tree T'. Let P be the predecessor of T' in the path. Since S initiates the path and T is closed, P is a member of T. Since P is the predecessor of T', there is an incentive to switch from P to T', so T' does not contain P. However, T' is a subtree of T since T is closed. Since every member of T appears in every closed subtree it immediately precedes, P appears in T'. This contradictory conclusion refutes the supposition that S initiates a terminating path.

The lemma holds alike for paths of nonrelative incentives and paths of relative incentives. Paths of relative incentives are paths in which incentives are relative to profiles. The nodes of the paths are strategies in the context of profiles rather than strategies in isolation. A *terminal profile* of such a path is any member of its terminal tree of profiles. The lemma asserts that no terminal profile starts a terminating path. The proof has the same structure as for paths of nonrelative incentives with strategies taken in isolation. Our existence proof for equilibria uses the version of the lemma for paths of relative incentives although we often speak of strategies as the nodes of paths, taking it as understood that we mean strategies in the context of profiles.

The proof of the existence of an equilibrium in every ideal normal-form game with a finite number of agents is by induction on the number of agents. For the basis step take the case of a one-person game. An equilibrium is a profile of strategies that are self-supporting given the profile. So in a one-person game an equilibrium is a strategy that is self-supporting. Hence to show that an equilibrium exists, it suffices to show that the agent has a strategy that does not initiate a terminating path. This result follows from the theorem of Section 4.3 that in every decision problem there is an option that does not start a terminating path. We can also obtain the same result a slightly different way using the lemma given previously. To find a strategy that does not start a terminating path, pick a strategy arbitrarily. If it is not itself a strategy that does not initiate a terminating path, pursue a terminating path of incentives it initiates to a terminal strategy, either a strategy that terminates the path or a strategy that is a member of a tree that terminates the path. According to the lemma, there is no terminating path of incentives away from the terminal strategy. Hence no strategy initiates a terminating path unless one does not. It is impossible for all strategies to initiate terminating paths.

In the inductive step we show that if every n-person game has an equilibrium, then every $(n + 1)$-person game also has an equilibrium. Grant the inductive hypothesis that every n-person game has an equilibrium. Then consider an arbitrary $(n + 1)$-person game. No profile in the $(n + 1)$-person game can fail to be an equilibrium unless some profile in it is an equilibrium.

To see this, take an arbitrary profile in that game. Suppose that it is not an equilibrium. Then there is an agent with a terminating path of relative incentives away from his strategy in the profile. Follow that path to a terminal strategy, either the last strategy in the terminating path or a strategy from the terminal tree of strategies. The other agents have a relative response to the agent's terminal strategy, and the agent predicts it.[5] His terminal strategy is self-supporting relative to the profile P consisting of that strategy and the response to it. The agent does not have a sufficient reason to switch from his part in P given the profile because his part is a terminal strategy for him. By the lemma, it does not initiate a terminating path of relative incentives for him.

Moreover, in an ideal game the predicted relative response of the other agents to his terminal strategy is an equilibrium of the n-person game obtained from the $(n + 1)$-person game by taking the terminal strategy of the agent to be fixed. By the inductive hypothesis there is an equilibrium in the n-person game. Also, by the idealization that the other agents choose strategies that are jointly self-supporting if possible, their response is an equilibrium of the n-person game if there is one. Hence their response is an equilibrium of the n-person game. Since the predicted response of the other agents is an equilibrium in the n-person game, none of the other agents has a sufficient incentive to switch from his part in P given the profile. The agents are prescient and given the profile foresee the strategy of the agent and the strategies of each other. Given that foreknowledge, none has a sufficient incentive to switch. Consequently, in the $(n + 1)$-person game neither the agent nor any of his opponents has a sufficient incentive to switch from his strategy in P given the profile. Thus all the strategies of P are self-supporting given the profile. So P is an equilibrium in the $(n + 1)$-person game. Given pursuit of sufficient incentives by all but one, pursuit of sufficient incentives by all is possible. This completes the inductive step of the proof.

Given the basis step and the inductive step, we conclude by mathematical induction that every ideal normal-form game with a finite number of agents has an equilibrium, more precisely, a strategic equilibrium.

Let me explain more fully the role in the proof of the appeal to the ideal, namely, the assumption that the games treated are ideal. The idealization that agents are jointly rational if possible is used to support the assumption that agents reach an equilibrium if one exists. This assumption has a crucial role

---

[5] If we apply the proof using agents' unreduced or partially reduced incentive structures, it is possible that some agents have endless paths of incentives. Nonetheless, for any strategy of any agent, a response of other agents exists, since the response is based on their actual pursuit of incentives, and hence their completely reduced incentive structures. They do not pursue incentives endlessly and hence do not make an endless series of adjustments to the agent's strategy.

in the inductive step, in which it is claimed that agents in the n-person game adopt an equilibrium. The assumption that agents are prescient also has a crucial role. It entails that an agent predicts the profile realized, whatever the profile is, and predicts the response of other agents to his strategy, whatever his strategy is. In consequence, an agent's incentives given a terminal strategy hold with respect to an accurate prediction of the response of other agents. His lack of a sufficient incentive to switch from the terminal strategy entails his lack of a sufficient incentive to switch from the terminal strategy given the realization of the profile consisting of the terminal strategy and the response of other agents. Since that response is an equilibrium of the n-person game obtained by fixing the agent's strategy, and since agents are aware of a profile given its realization, no agent has a sufficient incentive to switch strategies given that profile.

Let me also explain more fully one crucial step of the proof. Let P′ be the profile of the n-person game that is the predicted response to the agent's terminal strategy. It is an equilibrium of the n-person game. The proof claims that if the other agents lack sufficient incentives to switch from their parts in P′ given P′, they also lack sufficient incentives to switch from their parts in P given P. This holds although P is a profile of the $(n + 1)$-person game. The reason is that as far as their relative incentives are concerned, the supposition that P is the same as the supposition that P′. Realization of P is the same as the agent's realization of his part in P and the other agents' realization of P′. The only difference between P as an outcome of the $(n + 1)$-person game and P as an outcome of the n-person game concerns hypothetical cases. Whereas in the $(n + 1)$-person game the agent's strategy may respond to changes in the strategies of other agents, in the n-person game the agent's strategy is fixed and does not respond to changes in the strategies of other agents. However, since the game is normal-form, this difference does not affect the other agents' strategies. Moreover, since the game is ideal, this difference does not affect the other agents' relative incentives. Realization of a profile implies foreknowledge of it, so relative incentives go by payoff increases. These are the same for the other agents given P or given P′. Furthermore, in calculating the other agents' response to his strategy, the agent's strategy is fixed, just as it is in the n-person game. Although in the $(n + 1)$-person game other agents may predict some change in the agent's strategy that does not occur in the n-person game, their response to his terminal strategy in the $(n + 1)$-person game takes his strategy as fixed. A hypothetical change in the agent's strategy has a role only in the calculation of his response to their response. It does not have a role in determining their relative incentives given P.

The crucial point is that relative incentives hold with respect to a profile. Since the game is ideal, agents are prescient and have foreknowledge of the realization of a profile. So relative incentives are with respect to

157

foreknowledge of the profile supposed. As a result, in the profile P every strategy is self-supporting given the profile. The selected agent lacks a sufficient incentive to switch from his part in P since his strategy is a terminal strategy; given the response he predicts, he lacks a sufficient incentive to switch. By stipulation, P′ is the response he predicts and it is part of P. Hence given the profile P, he lacks a sufficient incentive to switch. P′ is an equilibrium for the other agents in the n-person game that results by taking the agent's terminal strategy as fixed. The other agents may not expect the agent's terminal strategy as his response to P′. The agent may have a better reply, one that other agents expect. But given the profile P, other agents know that he adopts the terminal strategy, and then they lack sufficient incentives to switch from their parts in P′ since it is an equilibrium in the n-person game that results from fixing the agent's terminal strategy. Hence given P, they lack sufficient incentives to switch. Because the equilibrium in the $(n + 1)$-person game is relative, it does not matter whether the strategies in it are self-supporting, period, and so self-supporting given expected responses. It only matters that they are self-supporting given the profile, and so self-supporting given the responses specified by the profile.

6

Finding Equilibria

Establishing the existence of equilibria is different from finding them. Section 5.3's proof of the existence of equilibria assumes the existence of the incentives and responses that determine equilibria but does not rest on a specification of them. Hence it does not show how to spot equilibria. It establishes the existence of equilibria in ideal normal-form games but does not provide a way of identifying them. This chapter formulates a procedure for finding equilibria in ideal normal-form games.

Let us begin with a few points about ideal games, our search procedure's subject. Although we have defined equilibrium in terms of incentives rather than payoff increases, in ideal games incentives follow payoff increases. Wherever an agent has an opportunity to increase his payoff, he has an incentive to switch strategy, and vice versa. This greatly facilitates identification of incentives and therefore equilibria. Also, in an ideal game agents are jointly rational if possible. Now that we have shown the existence of equilibria in ideal normal-form games, we also assume the existence of solutions. Therefore in the ideal games for which our search procedure is designed, we take the idealization about rationality to entail that agents are jointly rational. We also assume that their joint rationality carries over to their pursuit of incentives, which as a result obeys rules presented in Sections 6.2 and 6.3.

Our procedure for identifying equilibria applies to a restricted class of ideal normal-form games. The restrictions, related to the structure of incentives, are relatively mild. They are presented in the course of the chapter. To ensure that the procedure identifies strategic equilibria in a finite number of steps, it must also be restricted to games with a finite number of profiles. Restrictions are to be expected since search procedures are usually easier to formulate the more restricted the class of games to which they apply.

Even if we do not obtain a general procedure for finding equilibria, it is useful to have a procedure that works for some significant class of games. We would like a procedure for finding equilibria so that our equilibrium standard can help identify solutions to games, and so that we can support our equilibrium standard by verifying that the profiles it eliminates are intuitively not solutions. To make applications interesting, we need a procedure resourceful enough to find equilibria that are not Nash equilibria. We would like a procedure that enables us to compare the results of our equilibrium standard with the results of other equilibrium standards and other standards for solutions.

An important consideration in the design of a search procedure is the information available to find equilibria. Of course, a search procedure, if worthwhile, cannot require as input information specifying the equilibria of the game. It must start with information about incentives, or other more basic information about the game, and then use that information to find equilibria. Furthermore, since a search procedure has a practical objective, it should use information that yields equilibria in a simple way.

The equilibria of a concrete game, a payoff matrix's realization, are defined with respect to the agents' completely reduced incentive structures. The equilibria depend on the agents' information, not our information about the game. In an ideal game the agents know who abandons pursuit of incentives, and where pursuit of incentives is abandoned. Our information may not extend that far. Hence the incentives of agents and their paths of incentives may be different in the game and in the incentive structure given by our information. As a result, the profiles that are equilibria in the game are not necessarily the same as the profiles that are equilibria in that incentive structure. A search procedure relies on our information about the game rather than the agents' information. It computes equilibria in the incentive structure specified by our information. Our information has to include the pertinent information of the agents if the search procedure is to succeed. In ideal games agents have enough information, in virtue of idealizations about their knowledge, to locate equilibria for themselves. Our search procedure must have access to similar information.

We seek relative equilibria as opposed to nonrelative equilibria. In particular, we seek strategic equilibria, relative equilibria defined in terms of incentives. Strategic equilibria are profiles of strategies that are (strategically) self-supporting *relative* to the supposition that the profile is realized. To find relative equilibria, we need relative incentives and responses. Given an agent's switch from one profile to another, we need the relative response of other agents to the agent's part in the new profile and his relative incentives to switch given their response. To find relative equilibria, it is not enough to have information about pursuit of incentives with respect to the profile realized. We need information about pursuit of incentives with respect to every profile. We need to know the paths of incentives pursued by agents given the realization of any profile.

In specifying the information available to the search procedure, we go beyond the minimum needed to identify equilibria in ideal games. We include information that explains equilibria. Besides a specification of pursuit of incentives, we include a specification of the incentives pursued given relentless pursuit of incentives and the payoff matrix, which indicates all incentives pursued or not.

6.1. GROUP PATHS OF INCENTIVES

Our search procedure uses *group* paths of incentives as a heuristic device. They are artifacts constructed from agents' incentives by pretending that the group of agents in a game is an agent. Their introduction does not represent a change from an individualistic to a collectivistic perspective on equilibrium. My analysis of equilibrium remains individualistic – I still define an equilibrium in terms of agents' incentives. I do not define an equilibrium as a profile that fails to start a terminating group path. I do not assume that group preferences really exist. And I do not advance collectivistic explanations of the outcome of a game. I use group paths only to find equilibria taken as profiles of jointly, that is, conditionally, self-supporting strategies. I propose standards of rationality for group paths, but assume that those standards have an individualistic explanation.

This section describes group paths of incentives in detail. Group paths of incentives are used to define *group* equilibria. Our search procedure uses these artifacts to identify strategic equilibria. When we need a contrasting term for strategic equilibria, we call them *agent* equilibria. Our search procedure exploits a correspondence between group and agent equilibria. As Section 6.4 shows, we can use group equilibria to find agent equilibria.

A group path of incentives treats the group of agents as a single agent and displays the group's "incentives" to switch profile. On our view, a group has an incentive to switch profile just in case an agent has an incentive to switch his strategy in the profile. The group of agents may have incentives to switch from a profile to several different profiles. The incentive pursued depends on circumstances and principles of joint rationality. A possible pattern of pursuit of incentives designates for every profile at most one pursued incentive. A group path following incentives in a payoff matrix represents a possible pattern of pursuit of incentives by the group of agents.

Groups paths of incentives are similar to agent paths of relative incentives. Since they concern switches in profile, or group strategy, they have profiles as nodes, just as agent paths of relative incentives do. A group path of incentives follows agents' incentives, but whereas an agent path of incentives follows incentives of the same agent throughout, a group path may follow incentives of different agents in different segments. One segment may be an incentive for one agent, and another segment an incentive for another agent. Any incentive for any agent to switch from his strategy in a profile to another strategy may introduce a new profile. Then another agent's incentive to switch from his strategy in the new profile may introduce another profile, and so on. Two consecutive switches by the same agent are possible but may be prohibited by rules of rationality for pursuit of incentives. If an agent moves from strategy S1 to strategy Sn in $n - 1$ consecutive switches, the effect is the same as a single switch from S1 to Sn. In ideal games rules of rationality may require an agent to compress the series of steps into a single step; see Section 4.4. We leave the issue open but exclude consecutive switches by the same agent in our examples, unless entertaining criticisms.

In ideal games incentives correspond to payoff increases. An agent has an incentive to switch if and only if switching yields a payoff increase. Group paths of incentives follow agents' incentives and so follow payoff increases. In an ideal version of Matching Pennies, there are an incentive for the matcher, Row, to switch from (H, T) to (T, T), and an incentive for the mismatcher, Column, to switch from (T, T) to (T, H). So the following is a group path: (H, T), (T, T), (T, H). This path mixes Row's and Column's incentives.

Next, take an ideal normal-form game with the payoff matrix of Figure 6.1. Consider the profile (R1, C1). Row receives a payoff increase if he adopts R2 so that the profile (R2, C1) is realized. And then Column receives a payoff increase if she adopts C2 so that the profile (R2, C2) is realized. There is a group path of incentives from (R1, C1) to (R2, C2). It is (R1, C1), (R2, C1), (R2, C2).

Termination for a group path is defined the same way as termination for an agent path except that incentives for all agents are considered, not just for one agent. Specifically, a group path terminates in a profile if and only if no agent has an incentive to switch from his part in the profile. Realization of a profile, a group strategy, takes unanimity; a single agent suffices to block its realization. In the example of Figure 6.1, given (R2, C2) neither agent can switch strategies in a way that generates a payoff increase. So the group path presented terminates with (R2, C2).

A group path may have a tree of group paths as its last member. This happens if and only if an agent has an incentive to switch from the penultimate profile of the path to the first member of the tree, and the tree does not contain the profile. A group path that ends in a tree of group paths terminates in the tree if and only if the tree is closed and meets the minimality condition. The closure condition is met if and only if all incentives to switch away from members of the tree are represented by segments of the tree. The minimality

1, 1	1, 1
2, 1	1, 2

Figure 6.1 Group paths.

162

condition is met if and only if every member of the tree appears in every closed subtree it immediately precedes. Termination in a tree is the general case; it subsumes termination in a profile, a degenerate tree of group paths.

Agent and group equilibria exist with respect to an incentive structure. Group equilibria in an incentive structure depend indirectly on the incentives of agents in that incentive structure. Take an incentive structure for a game. An agent equilibrium with respect to it is a profile that does not initiate any terminating agent path in the structure. On the other hand, a group equilibrium with respect to it is a profile that does not initiate any terminating group path in the structure. In the example of Figure 6.1, (R2, C2) is a group equilibrium in the incentive structure of the payoff matrix since there are no incentives away from the profile in that incentive structure.

A group path for a concrete game, as opposed to a payoff matrix, indicates the way the group actually pursues incentives. An incentive to switch from a profile P1 to a profile P2 is pursued if and only if P2 is realized if the profile realized is either P1 or some profile to which some agent has an incentive to switch given P1. In one realization of the matrix of Figure 6.1, Row's incentive to switch from (R1, C1) to (R2, C1), the only incentive that exists given (R1, C1), is pursued. In that realization, given either (R1, C1) or (R2, C1), (R2, C1) occurs.

An agent may fail to pursue an incentive because he selects another, or because he resists the incentive despite having selected it, perhaps because other agents' pursuit of incentives preempts his pursuit of the incentive. Let us call an unpursued incentive of the first type an *unselected* incentive, and an unpursued incentive of the second type a *resisted* incentive. If an incentive is resisted because of its preemption by other agents' incentives, let us call it a *preempted* incentive too.

We distinguish selection of an incentive by an agent and by the group. An incentive selected by the group is selected by some agent. Each agent with an incentive given a profile has a selected incentive, but that incentive is not necessarily selected by the group. It may be preempted. An incentive selected by an agent but not by the group is preempted. A preempted incentive is selected by an agent but resisted by him. The group itself does not have preempted incentives, only resisted incentives. An agent's incentive selected by the group but not pursued is resisted not only by the agent but also by the group. An incentive an agent resists may be preempted, or it may be selected and resisted by the group. In a concrete game, group paths follow agents' incentives selected by the group and not resisted by it. Each segment of a group path is an incentive selected by an agent, not preempted by other agents' incentives, and not resisted by the agent.

We assume that pursuit of an incentive from one profile to another is independent of the profile supposed; that is, a conditional expressing pursuit

163

of the incentive is true even if embedded in suppositions of other profiles. Hence if the beginning of a group path of pursued incentives is pruned, the pattern of pursuit of incentives given by the remainder of the path is still accurate. Section 4.4 made a similar assumption about the independence of an agent's pursuit of incentives.

Since a group path for a game indicates the group's pattern of pursuit of incentives, it also indicates the relative response to an agent's strategy by the other agents. The group path (R1, C1), (R2, C1), for instance, indicates that in the context of (R1, C1), Row's response to C1 is R2. In a group path an agent switches from a strategy appearing in a profile only if his strategy is paired with the response to the strategy relative to the profile. A switch away from a strategy in a group path indicates that in the profile just before the switch the strategy is conjoined with the response to it given the profile. However, even without a switch away from a strategy, the strategy may appear with the relative response to it. This happens for a strategy in the terminal profile of a group path. The terminal profile itself contains the response to the strategy given the profile.

The contexts for agent and group incentives differ. Consider a group path. Take a profile in it. The group's incentive to switch profiles moves to the next profile in the path, and that profile sets the stage for further group incentives. In contrast, an agent's incentive to switch strategy given the profile moves to the next profile, the switch profile, and the subsequent response of other agents moves to the response profile for the strategy given the switch profile. The response profile, not the switch profile, sets the stage for the agent's further incentives to switch strategies.

To determine whether an agent has an incentive to switch strategies, group paths take agents' strategies in the context of a whole profile. The profile that sets the context for an agent's incentives is at first the initial profile of the group path. Afterward it is the profile that contains the other agents' response to the agent's part in a profile of the group path. In a group path a profile containing an agent's strategy may itself be the response profile for the agent's strategy in the context of the profile, but often changes in other agents' strategies take place before the response profile is reached. In other words, the relative response to a strategy given a profile is not necessarily the remainder of the profile. In fact, in some cases it is impossible for the relative response to be the remainder of the profile. The response to a strategy is the same relative to the profile introducing the strategy in the group path, and relative to every subsequent profile in the group path up to and including the response profile. If there are multiple profiles in this set, the response relative to each profile cannot be the remainder of each profile. It is the remainder of the response profile, not the remainders of the other profiles. For example, in a three-person game with the group path (R1, C1, M1),

(R2, C1, M1), (R2, C2, M1), (R2, C2, M2), the first two profiles are not the response profile for M1 in the context of (R1, C1, M1). Only the third profile, (R2, C2, M1), is the response profile, since it immediately precedes Matrix's switch from M1 to M2.

An agent's incentives relative to a noninitial profile of a group path are taken with respect to the response profile for his strategy in the profile. An agent's incentives relative to the initial profile are taken with respect to the initial profile itself. The initial profile of a multinode group path is not the response profile for all the strategies in the profile, however. Although the initial incentives of all agents are taken with respect to the initial profile, that profile has the response to only the strategy that changes in the first switch of the group path. The initial profile does not contain the responses to the other strategies.

Not every group path of a payoff matrix provides the response to an agent's strategy relative to a profile in the path. For example, in some group paths an agent's strategy in a profile is followed by profiles in which the strategies of other agents change endlessly. The response of other agents is thus unspecified by the path. But a group path for a concrete game avoids this result. Such paths do not continue endlessly. In general, the relative response to a strategy given a profile in a group path of a concrete game is the remainder of the profile if there is no subsequent profile, or if the subsequent profile contains a new strategy for the agent. If there is some subsequent profile with the same strategy, the relative response is the remainder of the last profile before a change in the strategy, or the remainder of the terminal profile if the path terminates in a profile before a change in the strategy.

Group paths indicate incentives pursued and responses. In ideal games, where pursuit of incentives is known, they also indicate beliefs about responses to strategies in profiles. Each step in a group path from one profile to another follows an agent's relative incentive to switch from one strategy to another given the strategies of other agents in the first profile as their relative response to the agent's strategy in that profile. The agent whose incentive to switch strategy changes the profile believes that the remainder of the profile is the response of other agents to his strategy in the context of that profile, and his incentive to switch strategies obtains with respect to that belief. Pursuit of incentives depends on knowledge of responses. Since responses follow pursuit of incentives, pursuit of incentives depends on knowledge of itself.

Next, let us consider an important difference between group and agent paths of incentives. In an agent's path of relative incentives, the agent's incentives are conditional on a profile, but in a group path of incentives, the group's incentives are not conditional on a profile. The objects of the group's incentives are entire profiles, not strategies forming parts of profiles.

Therefore, in contrast with the case for an agent, a profile is not needed to set the background for the group's incentives. The profiles that are the objects of the group's incentives themselves include all that is relevant.

This difference between group and agent paths affects the ease with which group and agent equilibria can be found. Since group incentives, in contrast with relative agent incentives, are independent of background profiles, it is easier to find group equilibria than agent equilibria. To follow a group path, one calculates for each profile in the path the incentives of agents to switch strategy given the profile. In contrast, to follow an agent path, one calculates for each noninitial profile of the path the response of other agents to the agent's strategy in the profile, and then the agent's incentives given their response. There is an extra step in following an agent's path – the calculation of other agents' responses. This extra step complicates the direct method of searching for agent equilibria. Search via group equilibria eliminates this extra step. It focuses on incentives without attending to responses. Responses are implicit in group paths, but we do not have to draw them out to find group equilibria. I identify agent equilibria using group equilibria because this is easier than identifying agent equilibria directly.

6.2. PARTIAL VERSIONS OF GAMES

This section introduces the *partial version* of a game, a specification of an incentive structure that explains the pattern of incentives actually pursued, the input for our procedure for identifying the game's equilibria. It describes partial versions, presents the features of games that ground partial versions, and examines various characteristics of and constraints on partial versions.

6.2.1. *Relentless Pursuit of Incentives*

If the group of agents in a game pursues incentives relentlessly, some incentive is pursued at every opportunity. For example, in an ideal version of Matching Pennies, given any profile, one agent pursues an incentive to switch strategy if pursuit of incentives is relentless. Hence pursuit of incentives is endless if relentless. Of course, in any realization of the game, one agent fails to pursue an incentive. Some profile is realized despite an agent's incentive to switch strategies given that profile. Although relentless pursuit of incentives is endless, actual pursuit of incentives stops at some profile.

A partial version of a game states for each opportunity to pursue incentives which incentive is pursued given the assumption that pursuit of incentives is relentless. If given some profile any agent has an incentive to switch, relentlessness implies that some agent pursues an incentive to switch. A game's partial version therefore indicates, among multiple incentives given a profile, the incentive pursued under the supposition that some incentive is pursued. If given a profile several agents have an incentive to switch, the partial version specifies the agent who pursues an incentive given relentlessness. If that agent has several incentives to switch, it also singles out the incentive the agent pursues given relentlessness. Each profile that generates an incentive generates exactly one incentive in the partial version. A partial version of Matching Pennies, for instance, specifies that agents pursue all incentives given relentlessness. Since at each profile there is just one incentive, none is discarded.[1]

Often a game's payoff matrix generates an elaborate tree of group paths away from a profile. For each agent who has an incentive to switch from his part in the profile, and for each strategy to which he has an incentive to switch, there is a path away from the profile. Each path that reaches a profile generating incentives sends out a branch for each incentive, and so on. A partial version of the game simplifies these trees of group paths. The simplification eliminates all but the incentives pursued given relentless pursuit of incentives. After the simplification, each profile initiates at most one path, not a tree of multiple group paths. Discarding unpursued incentives reduces a branching tree initiated by a profile to a single path. At most one path follows incentives pursued under the assumption of relentlessness. The group paths initiated by profiles then specify the incentive structure of the game's partial version. The partial version of Matching Pennies, for example, has a group path moving clockwise around the four corners of the payoff matrix. Hereafter in this section when we speak of group paths, we mean group paths in the incentive structure of a partial version.

A payoff matrix furnishes no information about the incentives that agents pursue when there are multiple incentives relative to a profile. This information is crucial for determining agents' paths of incentives and hence agent equilibria. An agent's incentives depend on his information about the incentives pursued by other agents, and his paths of incentives depend on his information about other agents' responses to changes in his strategy, that is, the incentives other agents pursue given those changes. A game's partial version provides information about pursuit of incentives. It informs by discarding some unpursued incentives, and so some unpursued group paths.

The incentives and group paths omitted by an ideal game's partial version have no influence on an agent's path of incentives; when an agent calculates other agents' responses to his strategies, he uses only the group paths they pursue. Consider an agent's strategy in a profile. Suppose that other agents

---

[1] Do not confuse a partial version of a game with the combination of the partially reduced incentive structures of agents. An agent's partially reduced incentive structure depends on the incentives actually pursued by other agents and not the incentives they pursue given relentlessness.

pursue incentives relative to the profile. Their pursuit of incentives gives their response to the agent's strategy in the context of the profile. In an ideal game any incentive of the agent to switch strategies holds with respect to their response. The partial version of an ideal game gives an agent's information about the direction of other agents' pursuit of incentives at profiles where multiple incentives exist. Thus it helps to calculate his incentives.

A group path has a tree as last member if and only if an agent's incentive goes from the tree's predecessor to the first profile in the tree and the tree does not contain its predecessor. The group path terminates in the tree if and only if the tree meets the closure and minimality conditions. Termination of a group path in a partial version is termination with respect to a pattern of pursuit of incentives, not with respect to all incentives. Hence closure of the tree is with respect to the incentives in the partial version. For the tree to be closed with respect to the partial version, it must include all incentives in the partial version generated by the tree's members. So the tree meets the closure condition with respect to the partial version if and only if for every profile P in the tree, if there is an incentive to switch from P to P' in the partial version, P' is in the tree too. To meet the minimality condition with respect to the partial version, every member of the tree must appear in every closed subtree it immediately precedes, where again closure is with respect to the partial version. In a partial version a group tree is a single group path of incentives selected for pursuit given relentlessness. Since in a partial version a profile starts at most one group path, closure in a partial version yields a path, not a branching tree. Hence a terminal tree of a group path is a subpath.

Let us examine closed group paths in a partial version, since they may form the termini of group paths. An infinite path, for every profile P in the path, includes as the next node the profile obtained by following the selected incentive away from P. This amounts to closure of the path. Hence a path is closed if and only if it is infinite or has a terminal profile as last member. A path is not closed if and only if it has a last member that is not a terminal profile. A closed path involving a finite number of profiles (which may appear in the path an infinite number of times) terminates in a profile or a cycle or is itself a cycle and so does not terminate in a subpath (because no closed subpath has a predecessor not in the subpath). The suppositions involved in paths of profiles are independent in the games we treat, so that an embedded supposition of a profile yields the same results as an initial supposition of the profile. Hence removing an initial segment of a group path does not cause any change in the remainder of the path. In particular, removing an initial segment of a closed path results in a closed path.

A partial version of a game gives the agents' pattern of pursuit of incentives given relentlessness. Selecting an incentive to be pursued given a profile in a normal-form game is similar to designating a player to move next in a multistage game. Pursuit of incentives in multistage games concerns the realization of profiles, and in single-stage games it concerns the realization of profiles given hypothetical, reduced sets of profiles. In either case, pursuit of incentives explains the nonrealization of profiles. Pursuit of incentives in normal-form games has a technical sense, however. Pursuit of an incentive relative to a profile entails that if the profile or some profile to which there is an incentive to switch is realized, the profile resulting from the incentive pursued is realized. In Matching Pennies, in which at each profile there is just one incentive to switch, an incentive to switch from P to P' is pursued if and only if given P or P', P'. Relentlessness entails that the incentive is pursued. This entails that given the realization of either P or P', the latter profile is realized. Relentlessness therefore generates endless pursuit of incentives. Under some nearness relations for possible worlds, the conditionals expressing pursuit of incentives may have indeterminate truth values. However, we assume that nearness among possible worlds is such that these conditionals have determinate truth values. By assumption, if some disjunction of profiles is realized, for one of them it is true that it is realized.

The consequences of relentless pursuit of incentives are problematic. Relentless pursuit of incentives moves along a group path to the terminal profile of the path if there is one. But consider the cycle of profiles connected by incentives to switch in an ideal version of Matching Pennies, namely, (R1, C1), (R1, C2), (R2, C2), (R2, C1), (R1, C1), ... It cannot be that for every ordered pair of adjacent profiles, if one profile in the pair is realized, it is the second. Some profile is realized, and the realization of any profile entails that in a context in which an incentive exists none is pursued. For instance, if (R1, C1) is realized, then Column does not pursue her incentive to switch to C2. In this case it is true that (R1, C1) or (R1, C2) is realized, and yet (R1, C2) is not realized. So it is false that if (R1, C1) or (R1, C2) is realized, (R1, C2) is realized. Therefore, the assumption of relentless pursuit of incentives is false.

The assumption's falsity is not itself a problem for the construction of a partial version. A model with a false assumption may have a useful explanatory role. But it also appears that relentless pursuit of incentives is impossible. If so, then assuming relentlessness does not yield a unique pattern of pursuit of incentives. Conditionals with impossible antecedents are either true or indeterminate regardless of their consequents. We cannot obtain a significant pattern of pursuit of incentives from impossible suppositions. We cannot draw useful conclusions about equilibria from the impossible.

To resolve the problem, we apply the assumption of relentless pursuit in a way that never requires us to draw a conclusion from an impossible

169

supposition. We take the assumption of relentlessness as attaching *pursuit of incentives* to each profile rather than as attaching *relentless pursuit of incentives* to each profile. We do not assume relentless pursuit of incentives along with the antecedent of a disjunctive conditional expressing pursuit of an incentive. Instead we assume pursuit of incentives along with the conditional's antecedent. To obtain the pattern of pursuit given relentlessness, we just assume pursuit at every opportunity. As a result, even if the overall pattern of pursuit of incentives is impossible, each piece of the pattern is possible and follows from assumptions that are possible.

In Matching Pennies, for example, we do not consider for each profile what happens if incentives are pursued relentlessly and either the profile or an incentive-reached alternative is realized. We consider for each profile what happens if incentives are pursued at that profile and either the profile or an incentive-reached alternative is realized. For each profile, we add only the supposition of pursuit of incentives to the antecedent of the conditional for that profile. As a result, for each profile the supposition is possible, even though for the profile realized the supposition is false. We never draw conclusions about pursuit of incentives from impossible assumptions.

Let me clarify this important point by putting it another way: According to our formulation of the assumption of relentlessness, *for each profile*, if under supposition of the profile any agent has an incentive to switch, it is assumed that some agent pursues an incentive to switch. The assumption of pursuit of an incentive is made *relative to each profile*. Under the assumption of relentlessness, in the disjunctive conditionals expressing pursuit of incentives, for each profile the supposition of pursuit of incentives is adjoined to the supposition that the profile or an incentive-reached alternative is realized. Under supposition of each profile, it is assumed that if there are any incentives to switch, one is pursued. For each profile, we thus obtain an assumption that is possible. The assumption for the profile realized and the conditional featuring that profile are in fact false, but they are possible.

We do not assume the conjunction of all such assumptions about pursuit of incentives for any inference about pursuit of an incentive. We do not assume that for all profiles if any agent has an incentive to switch, some agent pursues an incentive to switch. We do not use the global assumption of pursuit of incentives with respect to all profiles. That assumption is impossible in games such as Matching Pennies, in which at every profile some agent has an incentive to switch to another profile. In such games it is impossible that for all profiles if any agent has an incentive to switch, some agent pursues an incentive to switch. Some profile is realized despite an incentive to switch; pursuit of incentives is forgone at that profile.

The upshot of taking the assumption of relentless pursuit of incentives distributively is that the inferences made using the assumption rest on possible suppositions. Inferences about incentives are made singly from the distributed components of the assumption. We never have to make an inference about the direction of pursuit of incentives using an impossible supposition. Even if the combination of our conclusions about pursuit of incentives is impossible, each separate conclusion about an incentive is possible and stems from possible suppositions. Where the aggregate of the conclusions is impossible, one of the conclusions is false, and the corresponding component of the distributed assumption of relentlessness is false although not impossible.

There is another, related puzzle about the idealizations involved in partial versions. Take an ideal version of Matching Pennies again. Given the realization of a profile, some agent fails to pursue an incentive. The assumption of relentless pursuit of incentives and its apparent consequence, that each agent knows that pursuit of incentives is relentless, are both false. If each agent believes that he pursues incentives relentlessly, one agent is mistaken. If each agent believes that the other agent pursues incentives relentlessly, one agent is mistaken.

The assumption that pursuit of incentives is relentless creates an incompatibility of idealizations. In ideal games agents are prescient and fully informed about the game. They know the profile realized, the agents' incentives, and the agents' pattern of pursuit of incentives. But in an ideal game in which group paths of incentives are endless, prescience conflicts with knowledge of the game given relentless pursuit of incentives. Given relentlessness, it is impossible that each agent knows the profile realized, knows the other agents' incentives, and knows that the other agents pursue incentives at every opportunity. These assumptions are incompatible. In Matching Pennies none of the profiles is realizable given them. If each agent knows the profile realized and the other agent's incentives, then it is not the case that each agent knows that the other agent pursues incentives. If each agent knows, for example, that the profile realized is (H, T) and knows the other agent's incentives, then Column knows that Row does not pursue his incentive to switch to T. She cannot know that Row pursues all his incentives.

To work around the incompatibility of idealizations in partial versions, we give prescience priority over knowledge of the game. Where the idealizations conflict, we suspend knowledge of pursuit of incentives, not knowledge of the profile realized. When we assume a profile in a partial version, we suspend the assumption that each agent knows the other agents' pattern of pursuit of incentives and retain the assumption that each agent knows the profile realized, and so knows the other agents' choices. Knowledge of pursuit of incentives is subordinate to the idealization of prescience, the supposition of relentless pursuit of incentives, and the supposition that a profile is realized. We let knowledge of pursuit of incentives adjust to the other assumptions. For example, if in Matching Pennies (H, H) is assumed, we suppose that

171

each agent knows the other agent's strategy. Row has no incentive to switch given the profile since he knows the profile given its realization. Given relentlessness, Row believes that Column pursues her incentive to switch to T, but given the profile's realization, his belief is mistaken. Also, given the profile, Column believes that Row pursues his incentives. She concludes that pursuit of her incentive to switch to T is futile, since Row counteracts it. For the most part, we do not worry about the subtleties of agents' incentives in a partial version, since the emergence of the pattern of incentives actually pursued, the input for our search procedure, does not depend on agent paths of incentives with respect to a partial version, just group paths. We need only verify that under our interpretation of the assumptions used to derive a partial version of an ideal game, the assumptions are not inconsistent.

6.2.2. Nearest Alternative Profiles

An agent's path of incentives is the result of his choices in hypothetical situations. But groups do not choose, and so group paths are not the result of group choices in hypothetical situations. No collective psychological phenomena underlie group paths as individual choices in hypothetical situations and actual conditional preferences underlie agent paths. The group paths of a partial version are artifacts. Nonetheless, something real must underlie them if they are to generate a method of finding agent equilibria.

If collective choices and preferences do not underlie a group path in a partial version, what does? The conditionals expressing pursuit of incentives by agents do not underlie the path. First, those conditionals constitute the path. To explain the path, we have to explain them. Second, in some games not all of the conditionals are true. In Matching Pennies, for instance, the conditional for the profile realized is false. We need to derive those conditionals, and group paths of incentives, from a game's actual features. We may not simply posit a false pattern of pursuit of incentives if we hope to use the pattern to find and explain genuine equilibria of the game. We must therefore derive the partial version of a game from more basic features of the game together with the assumption of relentless pursuit of incentives.

Groups do realize profiles, even if they do not choose them, so conditionals concerning the agents' realization of profiles in hypothetical situations may serve as the basis of a partial version. A special type of conditional underlies group paths and the disjunctive conditionals expressing pursuit of incentives in a partial version. The underlying conditionals indicate the profile realized if a particular profile is *not* realized. These negative conditionals are the foundation for group paths and hence partial versions. They explain pursuit of incentives given relentlessness. The more basic features of a concrete game on which the derivation of its partial version relies are

172

expressed by these negative conditionals. In fact, the negative conditionals explain the direction of actual pursuit of incentives as well as the direction of relentless pursuit of incentives. As Section 6.3 shows, the negative conditionals provide a unified explanation of the agents' realization of profiles in both actual and hypothetical situations.

The negative conditionals concerning the realization of profiles form the reality behind relentless pursuit of incentives in games in which such pursuit is impossible in the aggregate. In such games we must derive the pattern of relentless pursuit of incentives from something actual. The negative conditionals, when combined with the distributed assumption of relentless pursuit of incentives, entail a particular pattern of pursuit of incentives. Where the pattern is impossible, the distributed assumption of relentless pursuit has a false component. But the negative conditionals are all true.

The pursuit of an incentive to switch from a profile P1 to a profile P2 is expressed by the conditional that P2 obtains if P1 or a profile reached by incentives from P1 is realized. Since an incentive to switch from P1 to P2 obtains with respect to P1, a fuller expression of pursuit of the incentive embeds the foregoing conditional in the assumption that P1. In other words, pursuit of an incentive to switch from P1 to P2 is expressed by the following conditional:

If P1 is realized, then if P1 or some profile to which there is an incentive to switch given P1 is realized, P2 is realized.

Such disjunctive conditionals give the result of the assumption of relentless pursuit of incentives. These conditionals are explained in terms of the negative conditionals mentioned. The negative conditionals, together with the assumption of relentless pursuit of incentives, yield the disjunctive conditionals that describe the course of relentless pursuit of incentives.

The negative conditionals on which the partial version of a game rests have the following form, where P1 and P2 are distinct profiles:

If P1 is realized, then if P1 is not realized, P2 is realized.

These conditionals purport to give the *nearest alternative* to a profile. They are similar in some ways to the disjunctive conditionals expressing pursuit of incentives. The negative conditionals are anchored by a profile, the profile in the antecedent. The disjunctive conditionals are also anchored by a profile, the profile in the antecedent, repeated as the first profile of the disjunction in the embedded antecedent; that profile yields the incentive-reachable alternatives that form the rest of the embedded antecedent. But the negative conditionals also differ significantly from the disjunctive conditionals expressing pursuit of incentives. The embedded antecedent of a negative conditional concerns the realization of profiles with respect to the set of all profiles save one, whereas the embedded antecedent of a disjunctive conditional concerns the realization of profiles with respect to a generally smaller subset of profiles. The subset of profiles involved in the negative conditionals is thus more nearly stable. Also, nearest alternative profiles are not grounds for the rejection of profiles. But pursuit of an incentive away from a profile indicates grounds, not necessarily conclusive, for rejecting the profile.

We must make one revision of the negative conditionals to take account of the special nature of counterfactual conditionals. Counterfactual conditionals interpreted in terms of nearness among possible worlds do not obey the law of contraposition. To express the intended exclusivity of the nearest alternative profile – it is to be the alternative realized if any alternative is realized – we must add the contrapositive of the embedded conditional. So in full detail the negative conditionals that interest us have this form:

If P1 is realized, then (if P1 is not realized, P2 is realized – and if P2 is not realized, P1 is realized).

The added conditional – if P2 is not realized, P1 is realized – entails every conditional with the same consequent and an antecedent entailing the added conditional's antecedent and compatible with its consequent. For example, suppose that P3 is distinct from P2. Then the added conditional entails that if P1 or P3 is realized, then P1 is realized. For among the not-P2-worlds are the P1-or-P3-worlds, so if the nearest not-P2-world is a P1-world, the nearest P1-or-P3-world is also a P1-world.

Suppose that for P1 the nearest alternative is P2. A consequence of the negative conditional that expresses this relationship is that with respect to P1 at most one incentive of at most one agent is pursued. If, for instance, agent A1 pursues an incentive from P1 to P2, then agent A2 does not pursue an incentive from P1 to P3. The pursuit conditionals for the two agents are (1) given P1, if P1 or some profile reachable from P1 via A1's incentives, then P2, and (2) given P1, if P1 or some profile reachable from P1 via A2's incentives, then P3. These conditionals are not incompatible because they have different embedded antecedents, but given the negative conditional and its contrapositive component they are incompatible.

Given a profile not all agents can pursue all of their incentives if multiple agents have an incentive or if some agent has multiple incentives. In an ideal game the incentive pursued, if some incentive is pursued, is indicated by the nearest alternative profile. The negative conditionals indicate the incentive pursued in the context of a profile if an incentive is pursued. If there is an incentive from P1 to P2 and some incentive away from P1 is pursued, then the incentive from P1 to P2 is pursued if and only if P2 is the nearest alternative to P1. This is the connection between nearest alternative profiles and pursuit of incentives in virtue of which nearest alternative profiles explain partial versions. The explanation of a partial version moves from nearest alternative profiles to pursuit of incentives given relentlessness.

The disjunctive conditionals expressing pursuit of incentives and group paths are explained by the negative conditionals expressing nearest alternative profiles. If a profile is not realized, its nearest alternative is. The assumption of relentless pursuit of incentives implies that the profile is not realized, assuming that some agent has an incentive to switch away from it. So if the profile or its nearest alternative is realized, its nearest alternative is realized given relentlessness. Likewise if the profile or some profile to which there is an incentive to switch is realized, the nearest alternative profile is realized given relentlessness – it is included among the profiles to which there is an incentive to switch. Nearest alternative profiles are the factors with which the assumption of relentlessness is conjoined to obtain the pattern of pursuit of incentives given relentlessness. Given relentlessness, the negative conditionals expressing nearest alternative profiles yield the disjunctive conditionals expressing pursuit of incentives. But the negative conditionals are not formulated under the assumption that pursuit of incentives is relentless. They are independent of that assumption. Consequently, supposition of a profile whose realization contravenes the assumption does not contravene those negative conditionals.

To elaborate, negative conditionals express nearest alternative profiles. P2 is the nearest alternative profile for P1 if and only if given that P1 is realized then (if P1 is not realized, P2 is realized – and if P2 is not realized, P1 is realized). Disjunctive conditionals express pursuit of incentives. The incentive from P1 to P2 is pursued if and only if given that P1 is realized if P1 or some profile to which there is an incentive to switch is realized, then P2 is realized. Pursuit of incentives given relentlessness is pursuit of incentives under a supposition. Its direction is expressed by the disjunctive conditionals under the supposition that for each profile an incentive is pursued if there is one.

A sequence of negative conditionals expressing nearest alternative profiles generates a sequence of disjunctive conditionals expressing pursuit of incentives given relentlessness and thus generates a group path. Take the sequence of conditionals: given P1 if not P1, then P2; given P2 if not P2, then P3; . . . ; given $Pn - 1$ if not $Pn - 1$, then Pn. The contrapositive components of the embedded consequents go without saying here. Combining the sequence of conditionals with the assumption of relentlessness, we obtain the sequence of conditionals: given P1 if P1 or an incentive-reached profile, then P2; given P2 if P2 or an incentive-reached profile, then P3; . . . ; given $Pn - 1$ if $Pn - 1$ or an incentive-reached profile, then Pn. Suppressing explicit relativization of incentives to profiles and adopting the convention

175

that adjacent profiles are connected by incentives, we obtain the group path: P1, P2, P3, ..., Pn − 1, Pn. The negative conditionals for a game, when combined with the assumption of relentless pursuit of incentives, yield the disjunctive conditionals and group paths.

Profiles can be arranged in a sequence so that the successor of each profile is its nearest alternative. We call the result a *sequence of nearest alternative profiles*. As we interpret nearness between profiles, sequences of nearest alternative profiles satisfy an independence requirement. The supposition of a profile in a sequence is embedded in the supposition of previous profiles. But those suppositions do not affect the profile's nearest alternative. Whatever the context, the nearest alternative to a profile is the same. In virtue of this independence assumption, if a profile P1 initiates a sequence S containing another profile P2, then the sequence initiated by P2 must be the subsequence of S that begins with P2's occurrence in S. Since a group path is a sequence of nearest alternative profiles, each profile in it is followed by the same sequence of profiles wherever the profile occurs. This grounds the independence assumption about group paths made earlier. Each occurrence of a profile in a path must be followed by the same profile, its nearest alternative.

We assume that every ideal game has a partial version given by negative conditionals expressing nearest alternative profiles. In other words, at every profile, if any incentive is pursued given relentlessness, it leads to a nearest alternative profile given by a negative conditional of the form specified. Since these conditionals limit pursuit of incentives at a profile to at most one incentive for at most one agent, they entail a sort of coordination of pursuit of incentives among agents. The sort of coordination involved seems plausible. The agents in a normal-form game do not communicate, but they do not act in isolation either. They achieve a sort of coordination in virtue of being in a game together. At most one agent can pursue an incentive at a profile, and he can pursue at most one incentive there. Pursuit of incentives in our technical sense concerns the *realization* of profiles in hypothetical situations. And the profile realized given one agent's choice depends on the other agents' choices. However, if the coordination involved in a system of nearest alternative profiles does not arise in every ideal game, we can work around the problem. In lieu of assuming, as we do, that all ideal games have a system of nearest alternative profiles giving pursuit of incentives, we can restrict our search procedure to games that do have such a system.

Relentless pursuit of incentives may be circular, as in Matching Pennies. Such circular pursuit of incentives is puzzling. Pursuit of incentives concerns the realization of profiles, but the realization of any profile in a cycle of pursuit conflicts with the cycle. Nearest alternative profiles resolve this puzzle concerning relentlessness. It is easy to understand how nearest alternative profiles may form a cycle. The realization of a profile in a cycle of nearest alternative profiles does not conflict with the cycle. The cycle generates a cycle of pursuit only if none of its profiles is realized. In games such as Matching Pennies, a cycle of nearest alternative profiles occurs. It generates circular pursuit of incentives when it is assumed that pursuit of incentives is relentless, but not otherwise. Although circular pursuit of incentives is impossible, it is a consequence of possibilities, namely, circular nearest alternative profiles, and distributive relentlessness. The impossible pattern of pursuit of incentives arises from nonactual components of distributive relentlessness. Instead of just positing an impossible pattern of pursuit of incentives for a partial version, we derive the pattern from actual features of the game – nearest alternative profiles. The crucial difference between the relevant negative and disjunctive conditionals is the following: In cases in which relentless pursuit of incentives is impossible, the disjunctive conditional for at least one profile is false. But the negative conditional for that profile is true. Thus the negative conditionals can explicate the disjunctive conditionals.

Our search procedure for equilibria uses group paths, and to make it work we restrict it to games in which nearest alternative profiles meet certain conditions. The idealizations for games bring nearest alternative profiles into alignment with reasons, and priority for reasons that are incentives brings nearest alternative profiles into alignment with incentives. Wherever there is an incentive, there is a nearest alternative profile. But it can still turn out that there is a nearest alternative profile where there is no incentive. Nearest alternative profiles handle tie breaking, and so can be unguided by incentives. Since group paths are composed of incentives, we therefore require that nearest alternative profiles follow incentives exclusively in the games we treat. Then nearest alternative profiles generate group paths. Since we restrict our search procedure to games in which all nearest alternative profiles track incentives, all sequences of nearest alternative profiles are group paths. We do not assume that a principle of rationality requires agents to stay put unless they have an incentive to move. But we restrict ourselves to games in which agents behave this way. The restriction makes it possible to infer agent paths of incentives from group paths of incentives.

Because a partial version is a description of a concrete game, the group paths of a partial version must meet consistency requirements, and in ideal games they must meet rationality or coherence requirements. One coherence requirement for group paths is provided by the *interagent selection rule*. It lays down a global constraint on jointly rational pursuit of incentives that applies to the transition from payoff matrix to partial version. The constraint is the following:

177

The partial version's incentive structure has no node where the selected incentive does not start a terminating group path, but an alternative incentive of the payoff matrix if substituted for the selected incentive does start a terminating group path.

The constraint entertains one change in incentive and holds the rest of the partial version constant. Put a slightly different way, it states that if an agent's incentive in a group path initiates a nonterminating group path, there are no rival unpursued incentives for any agent that initiate a terminating group path when substituted for the agent's incentive. The rule imposing the constraint gives priority to sufficient "group incentives." Compliance entails that in a partial version if a group path away from a profile does not terminate, then at each profile of the path no replacement incentive starts a group path that terminates.

The interagent selection rule is reasonable for agents in an ideal game. The rule requires agents jointly to acknowledge reasons with greater weight. If the selection rule is violated in some case, the agent whose sufficient incentive is preempted by an insufficient incentive is irrational. He irrationally resists an incentive. He does not use his weight in jostling with other agents to pursue his incentives. We observe, without presenting the proof, that the interagent selection rule can always be satisfied.

Keep in mind that the interagent selection rule is not assumed by our characterization of equilibria or our search procedure for equilibria. It is advanced for illustrative purposes only. It provides a working hypothesis used to flesh out examples. By assumption, pursuit of incentives is jointly rational in the games we treat. The interagent selection rule gives substance to this assumption. Also keep in mind that the interagent selection rule is not the only rule governing group paths in partial versions. In ideal games the group paths conform to all rules of rationality, including rules addressing strategic considerations.

6.3. COMPLETE VERSIONS OF GAMES

Our search procedure requires more information about a game than is provided by a game's partial version; the information it requires is contained in what we call its *complete version*. To information about incentives and pursuit of incentives given relentlessness, a game's complete version adds information about pursuit of incentives, period.[2] It puts aside a partial

---

[2] Some ideal games have payoff matrix representations that can be filled out to obtain just one complete version. This may be the case because there are few strategies, because some strategies strictly dominate others, or because strategies tie. In these cases the complete version presents no information about pursuit of incentives that cannot be derived from the payoff matrix and idealizations. Still the payoff matrix itself does not contain the derived information about pursuit of incentives. Similar points apply to partial versions that yield only one complete version given idealizations.

version's assumption that pursuit of incentives is relentless. It recognizes that in some ideal games, such as Matching Pennies, relentless pursuit of incentives is impossible, and some rational agent fails to pursue incentives. It says who fails to pursue incentives, and where he fails to pursue incentives. Every concrete ideal normal-form game has a unique complete version.[3] Let us take a closer look at complete versions and what they add to partial versions.

6.3.1. Realistic Pursuit of Incentives

A partial version assumes that pursuit of incentives is relentless. More precisely, it assumes that at every profile where some agent has an incentive to switch, some agent pursues an incentive to switch. This assumption is false in some games. In an ideal version of Matching Pennies, for instance, every possible outcome entails that an incentive is resisted, that is, selected for pursuit but nonetheless forgone. Some incentives that are pursued given relentlessness are in fact not pursued. The complete version of a game puts aside the assumption that pursuit of incentives is relentless. Whereas the partial version omits only unselected incentives or incentives unpursued given relentlessness, the complete version omits all unpursued incentives. It retains only the pursued incentives.

All group paths terminate in a profile in the complete version of a game with a finite number of agents. Realism stops all endless group paths of the game's partial version. Strategic reasoning may involve the supposition that the profile realized belongs to an endless path of the partial version. Realism demands that one profile of the endless path be realized given that supposition, even if the supposition is counterfactual. Under a counterfactual supposition that the profile realized comes from the path, there still is some profile at which the path stops. Endless pursuit of incentives cannot exist in the game's complete version, even under counterfactual suppositions about the profile realized.

Moreover, in an ideal game rationality reinforces realism's termination of group paths. Rational resistance of futile incentives stops the unending group paths of the partial version. The agents do not need to make every group path terminate in a profile in order to avoid futile pursuit of incentives. Termination of the group path leading to the profile realized is enough for that. But robustly rational agents avoid futile pursuit of incentives in every decision situation, even ones that arise from counterfactual suppositions. Rational agents do not pursue incentives endlessly, even given a

---

[3] Section 6.4 shows that every complete version has an equilibrium that is an equilibrium of the concrete game it represents. If every concrete nonideal game had a complete version, it would have an equilibrium too. Since some concrete nonideal games lack equilibria, they also lack complete versions.

counterfactual supposition that the profile realized comes from some endless group path of incentives. Each group path terminates in a profile because endless pursuit of incentives is futile. Even in a counterfactual case, some agent forgoes pursuit of an incentive so that an endless path stops. Such forbearance is rational given that pursuit of the incentive is futile. Although pursuit of incentives concerns dispositions, not actual behavior, an agent forgoes endless pursuit of incentives for good reason.

A game's complete version stops every nonterminating group path in the game's partial version. It also stops terminating group paths of the partial version that are endless, those that terminate in cycles. The members of a terminal cycle have nonterminating paths away from them in the partial version. When these paths end in the complete version, the paths terminating in the cycle terminate in a profile instead.

Omitting an incentive of a partial version does not entail omitting subsequent incentives in a group path the incentive starts. Omitting an incentive is not the same as truncating a path at the incentive. In the partial version for a finite normal-form game, unending group paths cycle. One break in the cycle is enough to make all subpaths end. In the partial version for a normal-form game in which some agents have an infinite number of strategies, an unending group path may not cycle. Then an infinite number of break points are needed to make all subpaths end. Take the game of picking the higher natural number. Since its unending group paths do not cycle, its complete version specifies an infinite number of unpursued incentives. If a group path of a partial version is endless, the complete version removes enough incentives to end the path and all its subpaths.

If a group path away from a profile terminates in a game's complete version, it terminates in a profile. There are no unending paths that can serve as terminal subpaths. Furthermore, because a complete version omits unpursued incentives, each closed group path terminates in a profile. A closed path terminates in a profile just in case it ends, and every group path of a complete version ends. The profile where a closed group path stops is its terminal profile. That profile contains the response to each strategy in it. If an agent has an incentive to switch given the response to his strategy, he does not pursue it. He resists the incentive. (For brevity, I often speak of "paths" instead of "closed paths," taking closure to be understood.)

If a profile initiates a group path in a complete version, it initiates only one, and that path terminates. Hence absence of a terminating group path away from a profile entails absence of a group path away from the profile. This fact simplifies the conditions for a profile's being a group equilibrium. A group equilibrium in a complete version is a profile that does not initiate a terminating group path. So it is also a profile that does not initiate any group path.

Since the games we treat are restricted so that nearness between profiles follows incentives, in a complete version, as in a partial version, sequences of nearest alternative profiles yield group paths. The only difference is that sequences of nearest alternative profiles combine with the assumption of relentless pursuit of incentives to yield group paths in a partial version, whereas they combine with accurate descriptions of hypothetical choices to yield group paths in a complete version. For instance, suppose that P2 is the nearest alternative to P1. Then there is an incentive from P1 to P2. Also suppose that given P1 if P1 or some incentive-reached alternative is realized, P1 is not realized. Then it follows in the games we treat that the incentive from P1 to P2 is pursued and is part of the complete version. The sequences of nearest alternative profiles are the common explanation of the partial and complete versions of a game.

On our view, the payoff matrix representation of an ideal concrete normal-form game generally does not provide enough information to identify the game's equilibria. The game's complete version, in contrast, provides all the requisite information. (Whether it provides enough information to identify solutions we leave an open question.) For our purposes, the complete version of an ideal game furnishes the canonical representation of the game.

The information given by a complete version of a game is in the possession of the agents if the game is ideal. In an ideal game the agents know the incentives actually pursued. This information is important for strategic reasoning and so is covered by the overall idealization for games of strategy. In particular, the idealization concerning the agents' information about the game gives them information about pursuit of incentives. Some predictions about the profile realized can be derived from this idealization, but the idealization of prescience is independent since prescience concerns prediction given hypothetical situations that do not arise in the course of pursuit of incentives.

In ideal games the agents can replicate the strategic reasoning of other agents, and in games such as Matching Pennies the agents know that it is impossible for each agent to pursue incentives. The complete version indicates which incentives are pursued. In a complete version an agent's path of incentives depends on whether given his strategy in a profile other agents actually pursue their incentives. Suppose that in Matching Pennies Column does not pursue her incentives. Row's path of incentives away from (T, H) terminates with Row adopting H, knowing that Column's response is H, a response that fails to pursue incentives. Since Column does not pursue her incentive to switch from H to T given (H, H), that profile does not start a group path in the complete version. In the complete version (H, H) does not start a path for Row or Column, although in the partial version, because of Column's incentive to switch to T, it starts a nonterminating group path of incentives.

181

We use group paths of the partial version to explain resisted incentives, that is, selected but forgone incentives. The main purpose of the partial version of a game is to provide the wherewithal for explaining the rationality of an agent's failure to pursue incentives of the partial version. Endless group paths of an ideal game's partial version show places where a rational agent may abandon pursuit of incentives. In the case of an endless, nonterminating group path, any place in the path is a rational stopping point. In the case of an endless group path terminating in a cycle, profiles in the cycle are rational stopping points. The *interagent stopping rule* holds that incentives at other places must be pursued. It imposes the following constraint:

No incentive of the partial version that starts a terminating group path in the partial version is omitted from the complete version.

The rule requires the complete version to omit no sufficient "group incentive" of the partial version. It allows stopping pursuit of incentives only at insufficient "group incentives." The transition from partial to complete version is made all at once, not in steps during which sufficient group incentives are recalculated in light of prior omissions. The rule is global since its input and output are entire incentive structures, the partial version and the complete version, respectively. If a group path of the complete version of an ideal game terminates in a profile, and in the partial version of the game an agent has an incentive to switch away from his part in the profile, there is exactly one such agent, and he resists his incentive. By the interagent stopping rule, his incentive initiates a nonterminating group path in the partial version.

In contrast with the interagent stopping rule, the personal stopping rule requires that no sufficient incentive of an agent's partially reduced incentive structure be omitted from his completely reduced incentive structure. The two rules are not interderivable since the partial version of a game does not give the partially reduced incentive structure of an agent. The latter structure does not assume relentless pursuit of incentives by other agents.

According to the interagent stopping rule, a group path's termination in a partial version provides a sufficient reason for agents to pursue the path's incentives up to the terminus (a profile or a cycle). Hence in an ideal game each agent has a sufficient reason to pursue his incentives in the path up to the terminus. The rule lays down a condition that the agents' pattern of pursuit of incentives must satisfy to be jointly rational, that is, so that each agent's pursuit of incentives is rational given the whole pattern. We assume that the rule has an individualistic justification in terms of the unreduced and partially reduced incentive structures of agents, the personal selection and stopping rules, and other individualistic rules of rational pursuit of incentives. But we do not attempt to provide such an individualistic justification of the rule. That is a project for future work.

182

According to the interagent stopping rule, termination in a cycle in a partial version yields termination in a profile of the cycle in the complete version. To see this, suppose that an endless path terminates in a cycle in the partial version. The interagent stopping rule prohibits stopping at a profile before the cycle, since such a profile starts a terminating group path in the partial version. Stopping must occur at a profile in the cycle. The place of termination must be in the terminus of the path in the partial version. Thus the terminating path in the partial version provides an explanation of the place of termination for the path in the complete version.

The interagent stopping rule can always be satisfied. Stopping endless group paths of a partial version never requires omitting an incentive that starts a terminating group path in the partial version. The proof is omitted. Since the interagent stopping rule can always be satisfied, we simplify our idealization for pursuit of incentives. We originally assumed that agents pursue incentives in a way that is jointly rational *if possible*. But given the satisfiability of the interagent selection and stopping rules we now assume that agents pursue incentives in a way that is jointly rational. There is more to jointly rational pursuit of incentives than compliance with the interagent selection and stopping rules. Additional strategic considerations are also relevant. But the satisfiability of the two rules makes a good case for the possibility of jointly rational pursuit of incentives. We assume that it is possible and achieved in ideal games.

Finally, consider a profile where a selected incentive is resisted and so is excluded from the complete version although it is part of the partial version. The partial version may discard other incentives relative to that profile. Those unselected incentives are not restored in the complete version despite the selected incentive's being resisted. We assume that any incentive not pursued given relentlessness is not pursued at all. The selected incentive is the incentive pursued if any is pursued. If the selected incentive is not pursued, none of the incentives it preempts is pursued in its place. Hence those preempted incentives are not part of the complete version.

One may wonder why selection of one agent's incentive preempts another agent's incentive, since in normal-form games agents' choices are causally independent. The answer is that pursuit of incentives is related to realization of profiles, even if in hypothetical situations, and this concerns all agents. Also, one may wonder why an agent may rationally fail to pursue an incentive preempted by another agent's resisted incentive. The interagent selection and stopping rules guarantee that an incentive preempted by a resisted incentive is an insufficient group incentive with respect to the partial version. Hence its nonpursuit meets an important standard of joint rationality, and hence individual rationality. We assume that a jointly rational pattern of pursuit of incentives may exclude preempted incentives.

We can postpone an individualistic justification of the interagent selection and stopping rules, and an exhaustive account of jointly rational pursuit of incentives, since our main results about equilibria and our search procedure do not rely on our points about jointly rational pursuit of incentives. Those points serve only to sketch an explanation of complete versions of games in examples. In particular, nonpursuit of incentives is partially explained by the payoff matrix, the partial version, and the interagent selection and stopping rules. Preempted incentives are partially explained by the payoff matrix and the selection rule, and resisted incentives are partially explained by the partial version and the stopping rule. Jointly rational pursuit of incentives can be used to explain equilibrium since conformity to the rules of jointly rational pursuit of incentives does not by itself entail equilibrium.

6.3.2. Agent Paths from Group Paths

Now we come to the point on which our search procedure rests: The group paths of an ideal game's complete version provide a means of deriving agents' paths in their completely reduced incentive structures. The derivation is possible because a group path represents agents' paths too. Every profile of a group path starts a path for an agent. The agent whose incentive leads a group path away from a profile has a path away from the profile. Since the incentive structure of the complete version and the set of completely reduced incentive structures of agents each have all and only the pursued incentives, compression of a group path of the complete version to obtain incentives for a single agent yields a path for the agent in his completely reduced incentive structure. The rest of this section explains the compression procedure. The next section shows that the compressed paths obtained, agent paths of the complete version, do correspond to paths of agents' completely reduced incentive structures.

An agent's incentives to switch from a strategy S to another strategy depend upon his beliefs about other agents' strategies if he adopts S. We call those strategies the response to S. The incentives of a complete version are incentives for an agent relative to a profile. Group paths in a complete version follow relative incentives. These relative incentives depend on relative responses to strategies. Since in ideal games relative responses follow relative incentives, the group paths of complete versions also indicate relative responses, and the beliefs of agents about relative responses to their strategies. Hence we can obtain a path of incentives for an agent by omitting the part of a group path that represents the other agents' incentives, the incentives that generate their responses to the agent's strategies.

Section 5.2 distinguishes incentives with respect to a profile from incentives with respect to the response profile for a strategy in the profile. A profile does not start a path for an agent in the complete version unless the agent has an incentive relative to the profile taken as an initial profile, and hence taken "as is" rather than after substitution of the other agents' response to the agent's strategy in the profile. An agent has no path away from a profile if the agent has no incentive to switch strategies with respect to the profile taken "as is." In a profile taken as an initial supposition, prescience provides knowledge of the profile, and that knowledge is the source of incentives. Incentives relative to a noninitial profile use responses to the agent's strategy obtained by supposing that other agents act on their incentives in the group path after the profile.

The difference between an agent's incentives relative to initial and noninitial profiles of group paths influences our procedure for obtaining paths of incentives for an agent by compressing group paths of a complete version. For the initial profile of a group path, agents' incentives are taken with respect to the initial profile rather than the response profile. Therefore, the compression procedure yields a path for an agent only if the group path begins with an incentive of the agent. Only in this case does the first profile of the group path double as a first profile in a path for the agent.

At most one agent has a path away from a profile in a complete version. Incentives of other agents with respect to the profile are preempted. The only agent path of the complete version initiated by a profile is the one that starts with the first incentive of the group path the profile initiates. That agent path is for the first agent of the group path. It comes from the group path by compression. To obtain it, profiles between switches of the group path's initial agent are omitted. The agent paths obtained by this sort of compression of the complete version's group paths are by definition *agent paths of the complete version*, and agent equilibria in the complete version are defined with respect to them.

Agent paths require responses, and group paths give responses in ideal games. A group path starts with an initial profile relative to which an agent has an incentive to switch strategy. The initial profile is followed by another profile with a new strategy for the agent, a switch profile, then other profiles containing the agent's new strategy, and, step by step, the relative response to it. After the response profile is reached, the path has a switch profile, the same profile except for a new strategy for the agent, a strategy to which he has an incentive to switch given the response profile.

We want to reduce a group path to a path for its initial agent. Our procedure is to begin by omitting all profiles except those that form pairs of adjacent profiles of the group path with different strategies for the initial agent. Then we omit one profile of each such pair. The first pair includes the initial profile of the group path. We keep that profile. Subsequent pairs consist of a response profile followed by a switch profile. We omit the

185

switch profile, in conformity with our convention to represent agent paths using response profiles instead of switch profiles. The convention is apt because the switch profile can be computed from the response profile given the incentive relative to the response profile that the agent pursues, whereas the response profile cannot be so easily computed from the switch profile. After the group path's compression, all profiles except the initial profile and the agent's response profiles are omitted.

Closure of a group path in the complete version entails closure of the path for the initial agent obtained from it by compression. Since the group path includes the pursued incentives of all agents, its compression includes the pursued incentives of the agent. All closed group paths of a complete version have a terminal profile. Since any incentive selected at the terminal profile is resisted, all agents resist incentives at that profile, including the initial agent. So the initial agent's closed path in the complete version stops where the group path does. The terminal profile of the group path is his last response profile. It contains the response to his terminal strategy and is the terminal profile of his closed path in the complete version.

In ideal games incentives correspond to payoff increases. Hence agent and group paths of incentives are paths of payoff increases. Agent paths of payoff increases are derived from group paths of payoff increases by compression, that is, by suppression of the payoff-increasing switches of all agents except one, and then by representation of those switches by the response profiles from which they come. An agent path of payoff increases is a series of strategy switches by an agent such that each switch produces a payoff increase for the agent on the assumption that the other agents respond to the agent's strategies in the particular payoff-increasing ways given by a group path of payoff increases. For each group path, there is an agent path beginning with the first switch of the group path. Take Matching Pennies, as presented in Figure 6.2. One group path proceeds as follows: (H, H), (H, T), (T, T), (T, H), (H, H), ... It gives payoff-increasing responses of Row and Column to each other. Corresponding to this group path is the following path of payoff increases for Column: (H, H), (T, T), (H, H), (T, T), ... Here switch profiles for Column in the group path have been omitted. Suppressing Row's strategies in the profiles that form the nodes of Column's path, we obtain the following path of strategies for Column: H, T, H, T, ... Except for suppressed nodes and strategies of other agents, an agent path of payoff

1, -1   -1, 1

-1, 1   1, -1

Figure 6.2   Matching Pennies.

186

increases is the same as the group path of payoff increases from which it comes by compression.

In ideal games, where paths of incentives are paths of payoff increases, a compressed path of incentives for an agent can be obtained as follows: Start with a group path of incentives that the agent initiates. Omit profiles in which the agent's strategy is the same as in the preceding and the subsequent profile. The result is a series of profiles arranged as follows: first, a profile; second, a switch profile that differs from the first only in the strategy of the agent and such that a switch from the first strategy for the agent to the second yields a payoff increase given the constant strategies of the other agents; third, a response profile in which the agent's strategy is the same as in the second profile and other agents' strategies change to yield their payoff-increasing responses to the agent's switch in strategy; fourth, a switch profile in which only the agent's strategy changes, and it changes so that the agent's payoff increases given the constant strategies of the other agents; and so on, with payoff-increasing changes in strategy for the agent alternating with payoff-increasing changes in strategy for the other agents. Finally, omit the switch profiles from this series of profiles to obtain a path of incentives for the agent.

6.4. A SEARCH PROCEDURE

Our procedure for finding equilibria in a concrete game is to find group equilibria in the game's complete version. We claim that a profile is an equilibrium in the game if and only if it is a group equilibrium in the game's complete version. An equilibrium in a concrete game is an agent equilibrium, specifically, an equilibrium with respect to the agents' completely reduced incentive structures. We have to show that a profile is an agent equilibrium of this sort in a concrete game if and only if it is a group equilibrium in the game's complete version.

A profile is a group equilibrium of a game's complete version if and only if it does not start a terminating group path. In a complete version a profile starts at most one group path, and any path it starts terminates. So a group equilibrium of the complete version is a profile that does not start any group path. We have to show that in order to be an agent equilibrium of a game, a profile must not initiate a group path in the game's complete version, and that if it does not, then it is an agent equilibrium of the game. To do this, we show (1) that a profile is a group equilibrium in the complete version if and only if it is an agent equilibrium in the complete version and (2) that it is an agent equilibrium in the complete version if and only if it is an agent equilibrium in the game.

Let us start with (1). We have to show that in a complete version being a group equilibrium is necessary and sufficient for being an agent equilibrium.

187

We show necessity by showing that in a complete version if a profile is a not a group equilibrium, then it is not an agent equilibrium. That is, we show that if a profile initiates a group path, the profile is not an agent equilibrium; some agent has a sufficient incentive to switch from his part in the profile. We show sufficiency by showing that in a complete version if a profile is a group equilibrium, then it is an agent equilibrium. That is, we show that if a profile does not initiate a group path, the profile is an agent equilibrium – no agent has a sufficient incentive to switch from his part in the profile.

To show necessity, suppose that a profile starts a group path in the complete version. Compress the group path to obtain the path for the first agent to switch in the group path. By definition, this compressed path is the agent's path in the complete version. It terminates in a profile since the group path does. Hence the profile starts a terminating agent path in the complete version. Even if the agent's path stops at a profile where the agent has a resisted incentive, his path still terminates in the complete version, where resisted incentives have been omitted. Hence the profile is not an agent equilibrium of the complete version.

To show sufficiency, suppose that a profile does not start a group path in the complete version. Then in the complete version no agent has an incentive to switch given the profile. Consequently, no agent has a path away from the profile, and so, a fortiori, no agent has a terminating path away from the profile. Hence the profile is an agent equilibrium of the complete version. Q.E.D.

We have shown that in a complete version of a game a profile is a group equilibrium if and only if it is an agent equilibrium. It remains to show (2), namely, that a profile is an agent equilibrium of a game's complete version if and only if it is an agent equilibrium of the game. To prepare for the demonstration, we prove as a lemma that a profile initiates a path for an agent in the complete version if and only if it initiates a path for the agent in his completely reduced incentive structure. This is easy to show. Suppose that a profile starts a path for an agent in the complete version. Then he is the initial agent of a group path of the complete version that starts with that profile. The first incentive of the group path is an incentive the agent pursues, and so it is part of his completely reduced incentive structure. The profile therefore starts a path for the agent in his completely reduced incentive structure. Conversely, if a profile starts a path for an agent in his completely reduced incentive structure, then it starts a group path in the complete version. Since the agent is the initial agent of the group path, the profile starts a path for the agent in the complete version.

The lemma makes it possible to show easily that a profile is an agent equilibrium in the complete version if and only if it is an agent equilibrium in the concrete game. To show the conditional from left to right, suppose

188

that a profile is an agent equilibrium in the complete version. Then there is no terminating agent path away from the profile in the complete version. Since all agent paths terminate in the complete version, there is no agent path away from the profile in the complete version. Since, by the lemma, a profile initiates a path for an agent in the complete version if and only if it initiates a path for the agent in his completely reduced incentive structure, there is no path away from the profile in the completely reduced incentive structures of agents. So there is no terminating, completely reduced agent path away from the profile. Hence, by definition, the profile is an agent equilibrium of the concrete game.

To show the conditional from right to left, suppose that a profile is an agent equilibrium in the concrete game. Then, by definition, it is an equilibrium with respect to the agents' completely reduced incentive structures. That is, there is no terminating completely reduced agent path away from it, and so no completely reduced agent path away from it. Since, by the lemma, a profile initiates a path for an agent in his completely reduced incentive structure if and only if it initiates a path for the agent in the complete version, there is no agent path away from the profile in the complete version. So there is no terminating agent path away from it in the complete version. Hence the profile is an agent equilibrium of the complete version. Q.E.D.

Let us summarize the foregoing results. The complete versions of games discard unpursued incentives. As a result, we can use group paths in complete versions to find agent equilibria in games. Complete versions yield by compression of group paths the completely reduced incentive structures of agents. Hence an equilibrium of a game is the same as a group equilibrium of its complete version. A profile is an agent equilibrium of a game if and only if it does not initiate a terminating path of relative incentives in any agent's completely reduced incentive structure. This happens if and only if it does not initiate a terminating agent path in the game's complete version. And this in turn happens if and only if it does not initiate a terminating group path in the game's complete version. Being a group equilibrium of the game's complete version is both necessary and sufficient for being an agent equilibrium of the game. In an ideal normal-form game with a finite number of agents that meets our restrictions about nearest alternative profiles, a profile is a group equilibrium in the game's complete version if and only if it is an agent equilibrium in the game.

Our search procedure for a game's equilibria is an identification procedure. It finds all the equilibria of the game. They are the profiles that do not initiate a terminating group path in the game's complete version. Since all group paths terminate in a complete version, the equilibria of a game are the profiles that do not initiate a group path in the game's complete version. The success of our search procedure stems from the derivability by compression of agents' paths in their completely reduced incentive structures from group paths in the complete version. We need not carry out the compression in applications of the search procedure, however. The main advantage of searching for equilibria using group paths is the dispensability of agents' paths in their completely reduced incentive structures.

The existence of group equilibria in a complete version is easily proved. In a complete version all terminating group paths terminate in a profile. Moreover, if a group path terminates in a profile, the terminal profile does not initiate a terminating group path. So no profile starts a terminating group path unless some profile fails to start a terminating group path. It is impossible for every profile to initiate a terminating group path. If one profile initiates a terminating group path, then some other profile does not. Every complete version therefore has at least one profile that does not initiate a terminating group path. That profile is a group equilibrium of the complete version.

This existence proof can be extended to equilibria in a concrete game using the connection between group equilibria of a complete version and agent equilibria of the concrete game. Since every complete version has a group equilibrium, every ideal normal-form concrete game with a finite number of agents that meets our restrictions has an agent equilibrium.

The foregoing proof of the existence of equilibria in a concrete game is simpler than Section 5.3's proof. The simplification is possible because of results in this chapter about sufficient conditions of equilibrium, and because of the restrictions imposed on games, in particular, the restriction concerning nearest alternative profiles. Section 5.3's proof uses first principles and covers all ideal normal-form games with a finite number of agents. Also, it applies to all agent incentive structures, not just completely reduced incentive structures. As noted in Section 5.3, it shows the existence of agent equilibria with respect to the agents' unreduced and partially reduced incentive structures, as well as the existence of equilibria with respect to their completely reduced incentive structures. The simple existence proof just given does not replace Section 5.3's proof but provides independent verification of it.[4]

---

[4] The simple existence proof provides a way to show the existence of equilibria in cooperative games since it does not depend on the assumptions of causal independence invoked by Section 5.3's proof for normal-form games. Details are left for another occasion.

7

Applications

Let us now apply our search procedure for strategic equilibria to various games, first the problem cases for Nash equilibrium presented in Section 3.3, then other cases that bring out the distinctive features of strategic equilibria. Although the standard of strategic equilibrium is much weaker than the standard of Nash equilibrium, it is not empty. It many games it disqualifies some profiles as solutions.

The applications serve mainly to illustrate strategic equilibrium and the search procedure. But they also provide a test of our equilibrium standard. No genuine solution should fail to meet the equilibrium standard. The standard should not disqualify any profile that counts as a solution on firm intuitive grounds or according to firm independent standards for solutions. The applications verify that our equilibrium standard rejects only nonsolutions.

7.1. RETURN TO THE PROBLEM CASES

Our equilibrium standard for solutions applies to a *concrete* game – in the case of a normal-form game, a realization of a payoff matrix. To apply the standard, we construct the complete version of a concrete game and use our search procedure to identify equilibria, assuming that the games treated meet the restrictions Chapter 6 imposes on applications of our search procedure. We typically construct examples starting with a payoff matrix for an ideal normal-form game. Then we introduce the game's partial version, and afterward its complete version, which together provide an account of pursuit of incentives. We construct examples so that group paths conform with the interagent selection and stopping rules, and we generally assume that they conform with all other rules of jointly rational pursuit of incentives. Unless noted otherwise, we assume that agents meet all rationality constraints.

The complete version of a game is the result of a process of reduction of incentive trees. The process takes each group tree of incentives in the payoff matrix, reduces it to a group path in the partial version, and then reduces the group path to a terminating group path in the complete version. More precisely, the process starts with incentives relative to profiles. These are payoff increases of the payoff matrix in an ideal game. They generate group trees of relative incentives. Then the process sorts the relative incentives to obtain the relative incentives pursued given relentlessness. These relative incentives form the incentives of the partial version. The group trees of relative incentives in the payoff matrix are pruned to yield the group paths of the partial version. Next, the process culls the relative incentives of the partial version to obtain the relative incentives that are pursued, period. These form the incentives of the complete version. Unending group paths of the partial version are interrupted to yield terminating group paths in the complete version.

Distinct complete versions of an ideal game may be constructed from a single normal-form game type, or payoff matrix. The complete versions differ according to which incentives are pursued, in particular, whose incentive is pursued when it is impossible for all agents to pursue their incentives.

To begin our illustrations, let us reconsider the games Section 3.3 presented as problems for Nash equilibrium. How does strategic equilibrium fare in those games? The first example is Matching Pennies. Let us construct a complete version of an ideal game of Matching Pennies in pure strategies. To begin, create a partial version from the game's payoff matrix (in Figure 3.1). Suppose that under the assumption that pursuit of incentives is relentless, the response to any strategy of Row, the matcher, is a strategy that does not match, and the response to any strategy of Column, the mismatcher, is a strategy that matches. Then consider the following sequence of profiles: (H, H), (H, T), (T, T), (T, H), (H, H), ... Each switch away from a strategy after the first switch is a response to a response to the strategy, and every time a strategy appears it is eventually paired with the response to it. We can take the sequence of profiles as a sequence of nearest alternative profiles. Then the sequence indicates, for the partial version, the group path away from every profile.

The group paths in the partial version arise from each agent's incentives to switch under the assumption that pursuit of incentives is relentless, and hence under the assumption that the other agent pursues his incentives. If incentives are pursued, then, given the idealizations, each agent has an incentive to switch from Heads to Tails, and from Tails to Heads. Each of an agent's strategies gives him evidence that the other agent has chosen in order to frustrate his strategy. Consequently, given each strategy he has an incentive to switch strategy. As a result, every profile has a group path away, and every closed group path away from a profile is infinite – each cycles through all profiles endlessly. It follows that no profile initiates a terminating group path. No group path away from a profile terminates in a subpath since every closed subpath contains all profiles, and so contains its predecessor.

The incentives of the partial version are with respect to the false assumption that pursuit of incentives is relentless, and the assumption's consequence, given the idealizations, that each agent is certain that the other agent pursues incentives. But the agents know it is impossible for each agent to pursue incentives at every opportunity. One agent must lose, and so fail to pursue an incentive to switch to a winning strategy. The partial version shows the futility of both agents' pursuing incentives. If the agents both pursue incentives, they each seek an upper hand they cannot both gain. It is rational for one to break off endless pursuit of incentives. Nothing in the payoff matrix for Matching Pennies determines who resists an incentive, however. That is determined by special features of the matrix's realization.

Let us suppose that nature makes some ideal agents resisters and others nonresisters or that background conditions determine the agents' dispositions to resist, perhaps by some random process. Imagine that Row is not a resister, but Column is. As a result, given any profile, Row pursues incentives and Column does not. In the concrete game, given the idealizations, both agents know who breaks off pursuit of incentives, and where. The transition from the partial to the complete version of the game takes account of this information. Given (H, H) Row has no incentive to switch from H, so that profile does not start a terminating path for him in the complete version. Given the profile, his choice of H does not provide evidence that his choice is frustrated by Column's response (now that the assumption of relentlessness is put aside). He has nonstrategic foreknowledge of Column's strategy. He does not infer her strategy from his own strategy, but in a way that relies on his knowledge that she is a resister.

Given (H, H) Column does have an incentive to switch from H to T, but she does not pursue it. She knows that Row pursues incentives to switch. If she switches from H to T, she knows that Row responds to frustrate her strategy, and the profile realized is (T, T). Given (T, T) she has an incentive to switch from T to H. In her partially reduced incentive structure (H, H) starts the following closed path of relative incentives: (H, H), [(H, T)] (T, T), [(T, H)] (H, H), ..., that is, suppressing switch profiles, (H, H), (T, T), (H, H), ..., or simplifying, H, T, H, ... This endless path does not terminate in a tree. The closed tree started by any of the path's strategies is a closed subpath that contains its predecessor. Column's not pursuing her incentive to switch from H to T given (H, H) is rational. Her stopping pursuit of incentives at (H, H) complies with the interagent stopping rule, since in the partial version the group path away from the profile does not terminate. Her stopping at (H, H) also complies with the personal stopping rule, since in her partially reduced incentive structure H starts a nonterminating path given (H, H).

The complete version prunes the incentives Column does not pursue, so in it she has no path of incentives away from (H, H). Since in the complete version neither agent has a path of incentives away from (H, H), that profile is an agent equilibrium of the complete version. Moreover, it is an equilibrium of the concrete game since neither agent has a path of relative incentives away from it in his completely reduced incentive structure. Row has no incentive to switch given (H, H), and Column's incentive to switch given (H, H) is unpursued and so absent from her completely reduced incentive structure. The status of (T, T) is similar. It also is an agent equilibrium of the complete version and an equilibrium of the concrete game. Furthermore, since there are no agent paths away from (H, H) and (T, T) in the complete version, there are no group paths away either. Both profiles are group equilibria of the complete version. On the other hand, (H, T) and (T, H) start terminating paths of relative incentives for Row in the complete version and in his completely reduced incentive structure and start a terminating group path in the complete version. Hence neither profile is an agent or group equilibrium in the complete version, and neither is an equilibrium of the concrete game. The group and agent equilibria of the complete version agree, and they agree with the equilibria of the concrete game. This confirms our search procedure.

It is possible that an agent pursues incentives given some profiles but not given other profiles. Perhaps a rule of rationality requires abandoning pursuit of no more incentives than necessary to avoid endless pursuit of incentives. This is an open issue. Consider a variation of our example in which Column pursues some incentives but not all. This may alter the equilibria of the game. Suppose, for instance, that Column fails to pursue only her incentive to switch from (H, H) to (H, T). Then (H, H) is the unique equilibrium. On the other hand, suppose that Column fails to pursue only her incentive to switch from (T, T) to (T, H). Then (T, T) is the unique equilibrium. If just one agent resists at just one point, only one profile is an equilibrium. Any profile can be the unique equilibrium depending on who resists where. In general, the equilibria depend on the details of the realization of Matching Pennies.

As the foregoing examples show, incentives away from an equilibrium of an ideal concrete game may exist, but if they do, agents have good excuses for not pursuing them. Suppose a profile is an equilibrium. Then according to our search procedure, there is no group path away from the profile in the complete version. Hence there is no pursued incentive away from the profile, at most resisted incentives. By the interagent stopping rule, a selected but resisted incentive is an insufficient group incentive in the partial version. By the interagent selection rule, the unselected incentives are also insufficient group incentives when substituted in the partial version. Agents need not pursue such insufficient incentives.

The complete version of a two- or three-person finite game in pure strategies may be represented by a box and arrow diagram. First, the group paths in the game's partial version are represented by arrows going from profile to

Figure 7.1   A partial version of Matching Pennies.

Figure 7.2   A complete version of Matching Pennies.

profile to indicate incentives selected for pursuit given relentlessness. Then the group paths in the game's complete version are represented by crossing out the arrows for selected but resisted incentives. In a game's complete version all group paths terminate. So in the diagram of a complete version, where paths of uncrossed arrows represent group paths, all paths of uncrossed arrows terminate. A group equilibrium of a complete version is a profile that does not start a terminating path of uncrossed arrows, or, more simply, a path of uncrossed arrows.

Figure 7.1 has the box and arrow diagram for Matching Pennies. Since there are only two options per agent, there is only one partial version. The arrows indicate the direction of pursuit of incentives given relentlessness. A complete version removes the assumption of relentless pursuit of incentives and specifies the selected incentives that are not pursued. In an ideal game, where agents are rational, the interagent stopping rule mandates that these be insufficient group incentives, ones that initiate nonterminating group paths in the partial version. In Matching Pennies all incentives of the partial version are insufficient group incentives. A complete version constructed from the partial version may specify any as unpursued. Suppose that Column does not pursue her incentive to switch from (H, H) to (H, T). In this case Figure 7.2 depicts the complete version. The crossed out arrow represents the incentive of the partial version that is not pursued and is omitted in the complete version. Crossed out arrows are not incentives of the complete version, but discarded incentives of the partial version. An incentive an agent resists does not appear in the complete version and is not part of any group path of the complete version. In the realization of Matching Pennies represented by Figure 7.2, (H, H) does not start a terminating group path of incentives in the complete version. It is an equilibrium of the concrete game

195

according to our search procedure. All other profiles start terminating group paths in the complete version and are not equilibria of the concrete game.

Next consider the game of choosing the higher number introduced by Figure 3.2. In this two-person game each agent has an infinite number of strategies. Since the game is ideal, given any profile each agent knows the profile, in particular, the strategy of the other agent. We are interested in the agents' incentives relative to a profile. They yield relative responses and relative equilibria. The first agent, for example, has an incentive to switch from 5 to 7 given the profile (5, 6), but not given the profile (5, 4). Under the idealizations, for each profile the agent with the lower number, or each agent if they have the same number, has an incentive to switch to a number higher than his opponent's number. Since there is an infinite supply of such numbers, he has an infinite supply of incentives to switch, each of which is optimal.

In a partial version, at each profile one incentive is designated as the incentive pursued if pursuit of incentives is relentless. In profiles that assign the same number to both agents, the designated incentive indicates the agent who pursues an incentive given relentlessness. In other profiles the designated incentive indicates the incentive pursued by the agent with the lower number given relentlessness. Take the profile (5, 5). The partial version may specify that the second agent pursues her incentive to switch to 6 given relentless pursuit of incentives. Next, take the profile (5, 6). The partial version may specify that among the numbers greater than 6 to which the first agent has an incentive to switch, he switches to 7 given relentless pursuit of incentives. In general, the agent a partial version designates to pursue an incentive given a profile may pursue his incentive to switch to a number greater by 1 than his opponent's number. In our example, we assume that agents behave this way, and that in case of ties the second agent pursues her incentive to switch. In the game's partial version every group path of incentives initiated by a profile is nonterminating. The group path starting with (5, 6) is (5, 6), (7, 6), (7, 8), (9, 8), ...

To find the game's equilibria, we have to put aside the assumption that pursuit of incentives is relentless. We need the game's complete version. Since relentless pursuit of incentives is impossible, the complete version puts a stop to it. The complete version stops all endless group paths of incentives. It recognizes that when the game of picking the higher number is over, one agent wins and the other loses. The loser fails to pursue incentives with respect to the profile realized. Since for each profile the group path it starts is nonterminating, pursuit of incentives may stop at any profile without violating the interagent stopping rule.

To stop all endless paths, it is not enough to make one agent stop pursuit of incentives at some number, for higher numbers still initiate endless paths for

196

him. If, say, the second agent pursues incentives up to 5 and stops, then, for her, paths of incentives initiated by lower numbers stop at 5. But closed paths of incentives initiated by higher numbers are still endless. To put a stop to every endless path of incentives, we need an infinite number of stopping points.

Suppose, then, that in the complete version one agent is a resister. One agent, say, the first, never pursues incentives to switch. This trait cuts off every group path of incentives. Every group path terminates in a profile in which the first agent has the lower number. It is no longer true that every profile initiates a group path. If in a profile the first agent has a lower number than the second agent, then the second agent has no incentive to switch. The first agent has an incentive to switch in the partial version. But he does not pursue it, and so has no incentive to switch in the complete version. So the profile does not initiate a group path in the complete version. If in a profile the first agent's number is the same as or higher than the second agent's, then the second agent moves higher than the first agent by 1, and the path stops. Every group path initiated by a profile terminates in a profile in which the first agent has the lower number. Accordingly, every profile in which the first agent's number is lower than the second agent's number is a group equilibrium. Our search procedure states that these group equilibria of the complete version are the equilibria of the game.

An appeal to first principles verifies the equilibria. The equilibria of the game by definition depend on paths of relative incentives in the agents' completely reduced incentive structures. In a profile in which the first agent has a lower number than the second agent, the second agent has no incentive to switch in the payoff matrix, and so none in her completely reduced incentive structure. The first agent does have an incentive to switch in his partially reduced incentive structure, in which it is assumed that he pursues incentives relentlessly. So in that structure he has a path away from his strategy. But the path does not terminate since the second agent pursues her incentives. If the first agent switches to a number higher than his opponent's, he knows then that his opponent has picked a number even higher. So he has an incentive to switch again. Since for each strategy in his path of relative incentives, he has an incentive to switch to another strategy, none of those strategies is a terminal strategy. Also, his path does not terminate in a subpath. The subpaths away from strategies lead to higher and higher numbers without end. So no closed subpath meets the minimality condition; every closed subpath has some member that does not appear in some closed subpath that it immediately precedes. For example, relative to the profile (5, 6) the first agent's path of incentives is (5, 6), [(7, 6)] (7, 8), [(9, 8)] (9, 10)..., or, simplifying by suppressing switch profiles and the second agent's strategies, 5, 7, 9, ... This path does not terminate in the subpath 7, 9, ... since 7 does not appear in 9, ...

197

Now put aside the assumption that the first agent's pursuit of incentives is relentless. Take a profile in which the first agent has a number lower than or equal to the second agent's number. He does not pursue his incentive to switch since he is a resister. Since the first agent's path away from the profile does not terminate in his partially reduced incentive structure, stopping pursuit of incentives at his number in the profile complies with the personal stopping rule. Since he does not pursue his incentive to switch, the profile does not initiate a path in his completely reduced incentive structure. The profile is therefore an equilibrium. Every profile in which the first agent has a lower number than the second is an equilibrium. Each such profile, if realized, has strategies that are self-supporting given the profile. No agent has a path, and a fortiori a terminating path, away from his strategy in the profile in his completely reduced incentive structure. The second agent has no incentive to switch in the payoff matrix, and so no incentive to switch in her completely reduced incentive structure. The first agent does have an incentive to switch in his partially reduced incentive structure, but he does not pursue it, and so it does not initiate a path of incentives in his completely reduced incentive structure.

The incentives we work with are relative to a profile. They are distinct from nonrelative incentives. An agent's nonrelative incentives to switch from a strategy assume a certain response that is not relative to a profile. In the preceding realization of the Higher Number Game, the first agent knows that the second agent picks the successor of any number he picks. The first agent has a nonrelative incentive to switch but does not pursue it. The responses on which the nonrelative incentives of the second agent are based have not been fully specified. Suppose the second agent picks a number. What response by the first agent does she anticipate? Since the first agent resists his incentives, it is a number lower than hers. She therefore has no nonrelative incentive to switch. By the idealization of prescience, the number she anticipates is in fact the lower number that the first agent selects. But we let unspecified features of the concrete game settle the lower number actually selected by the first agent. We do not need the number to calculate the relative incentives and relative equilibria that are our primary interest.

Next, let us apply our theory of equilibrium to Figure 3.3's game with an unattractive Nash equilibrium. Figure 7.3 repeats its payoff matrix. In the payoff matrix, there are multiple incentives to switch given some profiles. Given (R2, C1), for instance, Column has an incentive to switch from C1 to C2 and also to C3. The partial version of the game designates at each profile the incentive pursued given relentless pursuit of incentives. We suppose that the partial version is given by Figure 7.4. According to it, optimal incentives are selected. The box and arrow diagram gives a means of calculating the closed group path initiated by a profile. The closed group path initiated by

1, 1	1, 1	1, 1
1, 1	6, 5	5, 6
1, 1	5, 6	6, 5

Figure 7.3   An unattractive Nash equilibrium.

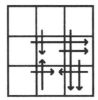

Figure 7.4   Pursuit of optimal incentives.

Figure 7.5   Resisted incentives removed.

(R2, C2), for example, cycles and so does not terminate. The cycle does not terminate in a closed subpath since every closed subpath includes its predecessor.

Now let us introduce a complete version of the game. We obtain one by selecting a profile at which an agent resists an incentive in the group cycle of incentives noted. Suppose that Column resists her incentive to switch from (R2, C2) to (R2, C3). Suppressing the arrow representing that incentive in Figure 7.4 yields the depiction of the game's complete version in Figure 7.5. Column's resisted incentive initiates a nonterminating group path of incentives in the partial version, so resisting that incentive complies with the interagent stopping rule. After we eliminate her resisted incentive to obtain the complete version, the group cycle of incentives stops at (R2, C2). The only profiles that do not initiate a terminating group path, a terminating path of arrows, are (R1, C1) and (R2, C2). According to our search procedure, those profiles are the group equilibria of the complete version and so are the equilibria of the game.

199

In this game the unique Nash equilibrium, (R1, C1), is a strategic equilibrium. But as a candidate for a solution, it competes with a rival strategic equilibrium, (R2, C2). (R2, C2) is more attractive as a solution than (R1, C1). As Figure 7.3 shows, it is strictly Pareto superior to (R1, C1), and its strategies, R2 and C2, dominate R1 and C1, respectively. (R2, C2) appears to be the solution of the game. The acknowledgment that two strategic equilibria exist accommodates the intuition that (R1, C1) is not the unique solution. It is a virtue of the standard of strategic equilibrium that it accommodates this intuition.

Our search procedure's identification of the strategic equilibria in the foregoing game is verified by first principles. That is, (R1, C1) and (R2, C2) are the strategic equilibria of the game according to the basic characterization of strategic equilibria in terms of agents' paths of relative incentives in their completely reduced incentive structures. To show this, we obtain the agents' completely reduced incentive structures from the complete version by compression of group paths. Only compression of a group path that starts with an agent's incentive yields a path for him. Neither agent has an incentive away from (R1, C1), so it is an equilibrium. In contrast, none of the other profiles with R1 or C1 is an equilibrium. Take (R1, C3), for example. Row has an incentive away from (R1, C3). To obtain Row's path away from (R1, C3), we compress the group path away from that profile. The group path is (R1, C3), (R3, C3), (R3, C2), (R2, C2). The terminal profile is both a switch and a response profile for Row. So Row's path is (R1, C3), (R3, C2), (R2, C2), or R1, R3, R2. It is a terminating path. Row's path from (R1, C3) to (R2, C2) in his completely reduced incentive structure terminates at (R2, C2) since Column fails to pursue her incentive to switch from C2 to C3. Therefore (R1, C3) is not an equilibrium.

When considering Row's incentives relative to (R3, C3), recall that his incentives depend on whether the profile is taken as an initial profile. See Section 5.1. Row has no incentives with respect to (R3, C3). If (R3, C3) is an initial profile, Row has no incentives relative to it. But when a profile containing an agent's strategy is not an initial profile, the existence of an incentive of the agent to switch from the strategy depends upon the existence of an incentive to switch given the response to the strategy, and not the existence of an incentive to switch given the remainder of the profile. Row's incentives given (R3, C3) are one thing, and his incentives given (R1, C3) and a switch to R3 are another. Row does have an incentive relative to (R3, C3) taken as a noninitial profile. If (R3, C3) results from Row's switch from R1 to R3, his incentives to switch from R3 are with respect to the response profile, (R3, C2). With respect to the response profile, he does have an incentive to switch to R2. This explains the second step in the condensed version of his path: R1, R3, R2.

For each of (R3, C2), (R2, C3), and (R3, C3), some agent has a terminating path of incentives away from the profile in his completely reduced incentive structure. For example, Row's path of incentives away from (R3, C2) terminates with (R2, C2) since Row has no incentive to switch from (R2, C2) given that Column does not pursue her incentive to switch from C2 to C3. Of the profiles in the cycle, only (R2, C2) is an equilibrium. Given (R2, C2), Row has no incentive to switch. Although Column does have an incentive to switch in her partially reduced incentive structure, she does not in her completely reduced incentive structure since she resists that incentive.

In the complete version of a game, every group path terminates and every terminating group path terminates in a profile. Likewise each agent path obtained by compression of a group path terminates and terminates in a profile. The same holds for paths in an agent's completely reduced incentive structure. An agent's path may terminate in an endless subpath in his partially reduced incentive structure, and then stop at a profile in the subpath in his completely reduced incentive structure. The endless but terminating path in his partially reduced incentive structure provides an explanation of the path's place of stoppage in his completely reduced incentive structure. According to the personal stopping rule, the place of stoppage must be in the terminal subpath of his partially reduced incentive structure so that no sufficient incentive is resisted. Our example illustrates this point, but we must identify partially reduced incentive structures to bring it out.

An agent's partially reduced incentive structure takes account of other agents' actual pursuit of incentives. It assumes relentless pursuit of incentives by the agent, but not by other agents. In our example Column's partially reduced incentive structure can be obtained by compression of group paths in the game's partial version because Row's pursuit of incentives is actually relentless. This compression procedure does not work in every case. The partial version does not yield the partially reduced incentive structure of Row, for example, since Column resists an incentive and does not pursue incentives relentlessly. Compressing the group path that starts at (R2, C3) in the partial version, we obtain for Row a path that does not terminate. But in his partially reduced incentive structure, the path starting with that profile terminates at (R2, C2).

To illustrate the preceding point about the personal stopping rule, we extract Column's paths in her partially reduced incentive structure from the game's partial version. Column's closed path away from (R3, C1) terminates in the cycle (R2, C2), (R3, C3), (R2, C2), ..., or C2, C3, C2, ..., in her partially reduced incentive structure. C1 does not appear in the cycle, and every member of the cycle appears in every closed subpath it immediately precedes. On the other hand, Column's closed path away from (R3, C2) is C3, C2, C3, ... This path does not end and does not terminate in a subpath.

201

Since the path forms a cycle, every profile in it is a member of every subpath the profile immediately precedes. In Column's completely reduced incentive structure her paths away from (R3, C1) and (R3, C2) both terminate in the profile (R2, C2) because Column stops pursuit of incentives at that profile. Halting there is authorized by the personal stopping rule.

7.2. ASSORTED EXAMPLES

To enrich our illustrations, let us consider additional examples making diverse points about strategic equilibria. The first example is a nonideal game in pure strategies with two agents. Both agents are prescient, but Column is irrational, and Row lacks direct information about the payoff matrix. Column knows the payoff matrix directly, but Row knows directly only that the payoff matrix is one of those in Figure 7.6. Row knows that Column is irrational and that she directly knows the correct matrix. Row is rational and Column knows this. Given Column's irrationality, Row knows that whatever profile is realized, Column's part in it is irrational. And given Row's rationality, Column knows that whatever profile is realized, Row's part in it is rational. Column's irrationality has priority over Column's predictive accuracy in case of conflict under counterfactual suppositions.

In these circumstances Column's choice carries information about the game for Row. That is, Row knows the correct matrix indirectly via foreknowledge of Column's choice. Suppose that a profile including C1 is realized. If Row assumes the first matrix is correct, then picking R2 is irrational, and if he picks R1, Column's choice is rational. Since Column's choice is irrational, the second matrix is correct. So the rational choice for Row is R2. He makes that choice, and given his choice Column's choice is irrational. On the other hand, suppose that a profile including C2 is realized. If Row assumes the second matrix is correct, then picking R1 is irrational and if he picks R2, Column's choice is rational. Since Column's choice is irrational, the first matrix is correct. So the rational choice for Row is R1. He makes that choice, and given his choice Column's choice is irrational.

Figure 7.6 presents the subjective game, our interest. The objective game is specified by Column's choice. It has a single complete version, whereas the subjective game does not. The subjective game has a distinct complete version relative to each of Column's choices, since supposition of a choice

1, 1	1, 0		0, 1	0, 0
		or		
0, 0	0, 1		1, 0	1, 1

Figure 7.6   Ignorance of the payoff matrix.

by Column affects knowledge of the payoff matrix and hence the incentive structure.[1]

To obtain a partial and also a complete version of the subjective game, it suffices to specify Column's strategy. Then there is at most one closed group path away from each profile, and all group paths terminate in a profile. The complete version's group paths fix the agent paths and equilibria. If Column does C2, Row knows the first matrix is correct. Since R1 is strictly dominant in that matrix, Row has a terminating path from R2 to R1 and no path away from R1. Column knows Row's response to C2 is R1 and so has a terminating path from C2 to C1, and no terminating path away from C1. So given C2, there is a terminating agent path away from every profile except (R1, C1). (R1, C1) is the equilibrium of the game. For similar reasons, if Column does C1, there is a terminating agent path away from every profile except (R2, C2). Then (R2, C2) is the equilibrium of the game.

The supposition of a strategy for Column settles the concrete game for which we calculate equilibria. We work out paths of incentives relative to profiles in that game, and not relative to the game that obtains if the profiles are realized. For example, if (R1, C1) is realized, then the game has the second matrix, and Row has a path away from (R1, C1). But supposing that Column does C2, the game has the first matrix. Consequently, Row has no path away from (R1, C1). In assessing Row's incentives, the supposition of (R1, C1) is embedded in the prior supposition of C2. The supposition of C2 introduces the nearest world w, where C2 is adopted, and the subsequent supposition of (R1, C1) introduces the world nearest to w where that profile is realized.

In the game of Figure 7.6, no matter whether C1 or C2 is realized, there is a terminating agent path away from every profile except one unrealized profile. If that profile is realized, there is still a terminating agent path away from every profile except one unrealized profile. Realization of a nonactual profile changes the incentive structure. As a result, the concrete game has no equilibrium. For every profile some agent has a sufficient incentive to switch given that profile; Column has an incentive starting a terminating path in her completely reduced incentive structure. The example shows that a nonideal game may lack an equilibrium and therefore a solution.

To defend strategic equilibrium against an objection, and to point out some open questions, we now turn to a certain case with multiple equilibria.

---

[1] A subjective game has a single complete version given our idealizations, or, more generally, given an incentive structure independent of the profile realized. A complete version is therefore a canonical version of an ideal subjective game, but not necessarily of a nonideal subjective game. On the other hand, every objective game has a single complete version. It is also the complete version of the corresponding subjective game when circumstances are ideal so that the objective and subjective games coincide.

2, 2	1, 1	-100, -100
1, 1	0, 0	1, 1
-100, -100	1, 1	2, 2

Figure 7.7   Multiple "unsafe" Nash equilibria.

The example is a symmetrical game of coordination in which agents are limited to pure strategies. There are two Nash equilibria, (R1, C1) and (R3, C3), and a failed attempt to realize one is penalized. See Figure 7.7. This type of game is sometimes used to argue against Nash equilibrium as a necessary condition for a solution. Since a failed attempt to coordinate is disastrous, it may seem wise to avoid the Nash strategies in favor of the "safe" strategies, R2 and C2. The Nash equilibria (R1, C1) and (R3, C3) seem dangerous given the agents' indifference between them. The safe course (R2, C2) seems rational although it is not a Nash equilibrium. This line of objection to Nash equilibrium is not forceful since it ignores the idealizations generally assumed for the standard of Nash equilibrium. If the game is ideal so that each agent is certain of his counterpart's collaboration in a particular Nash equilibrium, then the safe course loses its appeal.

Attention to background assumptions also defends our standard of strategic equilibrium in games with the payoff matrix of Figure 7.7. Suppose that circumstances are nonideal and each agent is uncertain of the other agent's responses to his strategies. Suppose that given the safe profile each agent assigns probability .5 to each of the other agent's Nash strategies. Then no agent has an incentive to switch. The safe strategies are jointly self-supporting, and the profile comprising them is a strategic equilibrium. On the other hand, in ideal circumstances, where agents are certain of the profile realized, the safe profile is not a strategic equilibrium. Given it, each agent has a sufficient incentive to switch to one of his Nash strategies. The Nash equilibria are the strategic equilibria since they alone fail to generate sufficient incentives to switch.

This example highlights some of the issues we reserve for future research. In an ideal version of the game, prediction of profiles requires foreknowledge of the results of breaking ties among equilibria. Our current explanation of prescience appeals directly to the agents' knowledge of the concrete game, including their tie-breaking behavior. We want a deeper explanation of prescience. In particular, we want to derive foreknowledge of the results of tie breaking from other features of ideal agents.

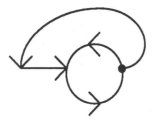

Figure 7.8   A preempted incentive.

Since the agents' rationality combined with other standard idealizations is insufficient for a derivation, we need novel idealizations concerning the agents' knowledge.[2]

Also, one wonders which of the equilibria are solutions. Given foresight of the equilibrium realized, is the other equilibrium a solution? Are its strategies rational given it? A standard of equilibrium provides only a necessary condition for solutions. Principles of equilibrium selection are needed to select solutions from among the equilibria. Does any principle of equilibrium selection apply to our example? The payoff matrix is symmetrical, so useful selection principles must appeal to the concrete game rather than the payoff matrix. They must take account of the way agents break ties and the like. Does any principle of rationality favor one of the equilibria given the agents' tie-breaking behavior and other details of the concrete game, or is each equilibrium a solution of the concrete game? Intuition suggests that both equilibria are solutions. But does this intuition address the abstract payoff matrix rather than a concrete realization of the payoff matrix, including the agents' tie-breaking behavior? Answering these questions about solutions requires an exploration of reasons beyond incentive structures, an area outside our purview.

The next example shows how equilibrium depends on actual pursuit of incentives. Imagine a game whose payoff matrix generates the incentive structure depicted in Figure 7.8. Also suppose the following features: The profile at the tail of the arrow starting on the left is not a group equilibrium in the game's complete version. The group path it starts terminates in a cycle in the game's partial version and terminates at a profile of the cycle in the game's complete version, the profile represented by the dot. The agent who resists the incentive selected at the terminal profile is not the group path's initial agent. The initial agent has a preempted incentive at the terminal

[2] See Sobel (1994, Chps. 13, 14) on "hyperrational" agents for views about what follows from the agents' rationality alone.

Figure 7.9   A preempted incentive in a three-person game.

profile. His preempted incentive leads him back to the group path's initial profile given other agents' pursuit of incentives. In his unreduced incentive structure the path the initial profile starts does not terminate. Is that profile an equilibrium of the concrete game?

For vividness, let us suppose that the preceding situation arises in a three-person game partially represented by Figure 7.9, where the first matrix of rows and columns appears in the foreground. The initial profile of the relevant group path is (R1, C1, M2), that path's initial agent is Matrix, and its terminal profile in the game's complete version is (R2, C2, M1). Is the initial profile (R1, C1, M2) an equilibrium of the concrete game? Only Matrix pursues an incentive away from that profile. The other agents have at most unselected incentives away. Matrix has a nonterminating path away from (R1, C1, M2) in his unreduced incentive structure. His path away returns via the profile (R2, C2, M1). His completely reduced incentive structure, however, takes account of his information about incentives he does not pursue. In that structure he has a terminating path away from (R1, C1, M2). That path terminates where the group path does in the game's complete version, at (R2, C2, M1). According to first principles, and also our search procedure, the group path's initial profile (R1, C1, M2) is not an equilibrium of the concrete game. Incentives away from a profile prevent equilibrium, despite starting paths back, if those paths include unpursued incentives.

Our final two examples illustrate equilibrium's sensitivity to the profiles available. They show that deletion of profiles may change remaining profiles from non-equilibria to equilibria. They explain why the standard of independence from irrelevant alternatives, commonly advanced for cooperative games, especially bargaining games, is inappropriate for normal-form games.

Let us begin with a statement of the standard of independence from irrelevant alternatives. The standard does not advance a necessary condition for a solution to a game, but rather a requirement of coherence for solutions to pairs of games. It states that if a profile is a solution to a game, then it is a solution to a reduction of the game where it remains feasible. In other

3, 3   0, 0

2, 2   2, 2

4, 0   0, 1

Figure 7.10   The standard of independence.

words, a reduction that retains a solution maintains the solution. Conversely, addition of profiles does not change a nonsolution to a solution.[3]

The standard of independence is not proposed for normal-form games because it conflicts with common views about solutions. Take an ideal normal-form game in pure strategies with the payoff matrix of Figure 7.10. In this game (R2, C2) is the unique Nash equilibrium. It is commonly regarded as the unique solution. But if R3 and the profiles containing that strategy are eliminated, then (R1, C1) is also a Nash equilibrium and is strictly Pareto superior to (R2, C2). (R1, C1) is commonly regarded as the unique solution to the reduced game. Granting the common views about solutions, the case violates the standard of independence. (R2, C2) is a solution in the original game, but not in the reduced game, even though it is still available.

Strategic equilibria coincide with Nash equilibria in this example. In a realization of the payoff matrix of Figure 7.10, there is just one strategic equilibrium, (R2, C2). So it is the solution. However, suppose R3 is eliminated. Then in a realization of the reduced payoff matrix, (R1, C1) is also a strategic equilibrium. Moreover, (R1, C1) is strictly Pareto superior to (R2, C2). So it is the unique solution of the reduced game, assuming the Pareto standard of equilibrium selection. Contrary to the standard of independence, a solution to the original game is not a solution to the reduced game despite its availability.

The next counterexample to the standard of independence dispenses with principles of equilibrium selection. It identifies solutions in a game and its reduction by finding the unique strategic equilibrium of those games. We imagine a two-person game whose payoff matrix generates the unreduced incentive structure depicted by Figure 7.11. Some profiles generate multiple incentives. Some generate two incentives for the same agent, and some generate an incentive for both agents. The partial version selects at most one incentive at each profile. Figure 7.12 presents the game's partial and complete versions. The crossed out arrow indicates an incentive

[3] Luce and Raiffa (1957: 350) express the independence standard for n-person bargaining problems this way: "Adding new (trading) alternatives, with the status quo point kept fixed, shall not change an old alternative from a non-solution to a solution."

207

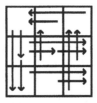

Figure 7.11   The incentive structure of the payoff matrix.

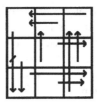

Figure 7.12   The partial and complete versions.

Figure 7.13   The reduction's partial and complete versions.

of the partial version discarded by the complete version. Every profile except (R1, C1) starts a terminating group path in the complete version, and hence that profile is the unique strategic equilibrium and solution of the game.

The reduced game is obtained by deleting R3 and C3. The reduced game's payoff matrix comes directly from the original game's payoff matrix. Figure 7.13 presents the reduced game's partial and complete versions. Some incentives selected in the reduced game are different from those selected in the original game since not all incentives selected in the original game are part of the reduced game. The selected incentive resisted in the reduced game is also different from the selected incentive resisted in the original game since the latter is not part of the reduced game. (R2, C2) is the reduced game's unique strategic equilibrium and solution. All other profiles start terminating group paths.

In this case the profile that is the original game's solution is not the reduced game's solution although it is feasible in the reduced game. The case is therefore a counterexample to the standard of independence. It is an

208

especially compelling counterexample since it does not assume a standard of equilibrium selection such as Pareto optimality.

The standard of independence conflicts with intuitions about solutions because elimination of profiles may convert a non-equilibrium into an equilibrium. Strategic equilibria are especially sensitive. Suppose elimination of profiles removes a strategy for an agent in a terminating path of incentives for the agent. Removal of the agent's strategy disrupts the terminating path and may make the path's initial strategy self-supporting with respect to the profile serving as its context. That profile may then be a new strategic equilibrium.

The standard of independence is attractive, so it is worthwhile to explain where the reasoning behind it fails, and why the variability in strategic equilibria it forbids is in fact acceptable. The main argument for the standard of independence for normal-form games relies on decision principle alpha, presented in Section 4.5. Principle alpha states that an option that is a rational choice in one decision problem is a rational choice in another decision problem obtained from the first by deletion of other options, provided the grounds of preference among options preserved do not change. Sometimes statements of the principle omit the proviso, but Section 4.5 shows its importance. Even with the proviso, principle alpha is false. It conflicts with the principle of self-support and should be rejected, as Section 4.5 argues. We let that pass in order to explore the reasoning behind the standard of independence.

The argument from principle alpha to the independence standard for solutions goes as follows:

Suppose that the standard of independence is violated. A profile P that is a solution in a game is not a solution in a reduction in which it is feasible. Now a solution is a profile of choices that are rational given the profile. So given P some option that is a rational choice in some agent's original decision problem is not a rational choice in the agent's reduced decision problem even though it is available. (The agent's "reduced" problem may be the same as his original problem if the game's reduction does not involve his strategies. Principle alpha still governs this degenerate case.) Since such a violation of principle alpha is prohibited, the violation of the standard of independence is also prohibited.

One problem with this argument is that a game's reduction often violates principle alpha's proviso that a reduction in options not change the grounds of preference among remaining options. In a normal-form game a reduction in profiles reduces an agent's set of strategies. If the game is ideal, then, because of strategic considerations, other agents may act differently. This alters the agent's information about their behavior and consequently changes the grounds of his preferences among strategies. For instance, in the example of Figure 7.10 Row has no incentive to switch from (R1, C1) once R3 is removed, and Column knows this. Column may then prefer C1 to C2. And in consequence Row may prefer R1 to R2. The assumption of fewer strategies for Row carries information about Column's strategies. Row originally assigns probability 0 to C1, but after the deletion of R3 may assign probability 1 to C1. There is a change in information about the states of the world Row considers in deciding what to do, and therefore a change in the grounds of preference among his strategies. Since information about other agents' strategies is relevant to an agent's preferences among his strategies, principle alpha cannot be used to support the standard of independence for solutions.

8

Other Standards for Solutions

This chapter compares the standard of strategic equilibrium with the two most commonly proposed standards for solutions: the standard of Nash equilibrium and the standard of nondomination. It also compares the standard of strategic equilibrium with other revisions of the standard of Nash equilibrium. All standards are taken to advance necessary conditions for solutions and to apply to ideal normal-form games, as the standard of strategic equilibrium does. This chapter's goal, in contrast with the preceding chapter's goal, is general comparison of the standard of strategic equilibrium with other standards for solutions, and not comparison of the standard's consequences with intuitions about solutions in particular cases. The chapter looks especially for conflict between the standard of strategic equilibrium and other standards. Are there cases in which it is possible to satisfy the standard of strategic equilibrium and possible to satisfy another standard, but impossible to satisfy both standards, so that the other standard argues against the standard of strategic equilibrium?

8.1. STRATEGIC AND NASH EQUILIBRIUM

Let us compare the standard of Nash equilibrium with the standard of strategic equilibrium, taking the two standards as potential rivals. We will see that the two standards do not conflict in ideal normal-form games. Where it is possible to satisfy each, it is possible to satisfy both, since every Nash equilibrium is a strategic equilibrium.

Not every strategic equilibrium is a Nash equilibrium. For example, some games have Nash equilibria and strategic equilibria but have more strategic equilibria than Nash equilibria. The game with an unattractive Nash equilibrium presented by Figure 7.3 is an illustration. It has a unique Nash equilibrium, but two strategic equilibria. The standards of strategic and Nash equilibrium do not conflict in that example, however. It is possible to satisfy both since the unique Nash equilibrium is also a strategic equilibrium. Realizing the Nash equilibrium thus satisfies both standards.

The example of Figure 7.3 raises the following question: Where some strategic equilibrium is not a Nash equilibrium, are there always multiple strategic equilibria? If there is only one strategic equilibrium, must it be a Nash equilibrium? Can we show that meeting the standard of strategic equilibrium entails meeting the standard of Nash equilibrium in games with a unique strategic equilibrium? One may reason as follows:

If a strategic equilibrium initiates a group path, the path does not terminate. And if a profile initiates a group path that does not terminate, then every profile in the path is a strategic equilibrium. No profile in a nonterminating group path can initiate a terminating group path, or the original profile would also initiate a terminating group path. So a strategic equilibrium is not unique if it starts a group path. A strategic equilibrium is unique only if it initiates no group path and thus is a Nash equilibrium.

This argument's claim about strategic and Nash equilibria holds for partial versions, but not for complete versions. In a complete version a profile can fail to initiate a group path, and so fail to initiate a terminating group path, and yet not be a Nash equilibrium, even if there is only one strategic equilibrium. See, for instance, the complete version of Matching Pennies in Figure 7.2. It has only one strategic equilibrium, and that profile is not a Nash equilibrium. Since the strategic equilibria of the game and its complete version coincide, this example shows that a game may have a unique strategic equilibrium that is not a Nash equilibrium.

To test the standard of strategic equilibrium, let us now look for games in which the standard conflicts with the standard of Nash equilibrium: that is, games in which both types of equilibrium exist but do not overlap, so that compliance with one standard entails noncompliance with the other standard. In other words, let us look for games in which it is possible to satisfy each standard separately but not possible to satisfy both standards together: that is, games in which each standard is met by some profile, but no profile meets both standards. If games of this type exist, defending the standard of strategic equilibrium requires arguing that none of the Nash equilibria is a solution. In a game with a Nash equilibrium and a strategic equilibrium but without a profile that is both, we must argue that a solution is a strategic equilibrium rather than a Nash equilibrium. We must argue that even where Nash equilibria exist, none is a solution if none is a strategic equilibrium.

The conflict we are seeking does not arise in ideal normal-form games. In them, although not every strategic equilibrium is a Nash equilibrium, every Nash equilibrium is a strategic equilibrium. If a profile is a Nash equilibrium, and so generates no incentives to switch in the incentive structure of the payoff matrix, it generates none in the incentive structures of agents. If no agent has an incentive to switch from his part in the profile, then no agent has a path, and a fortiori a terminating path, of incentives away from his part in the profile. It is a strategic equilibrium.[1]

As the foregoing shows, the standard of strategic equilibrium is merely a weakening of the standard of Nash equilibrium in ideal normal-form games. Keeping this in mind aids identification of strategic equilibria and solutions. For example, if there is just one Nash equilibrium, that profile is also a strategic equilibrium and, according to the standard of strategic equilibrium, is the solution if it is the unique strategic equilibrium.

8.2. ALTERNATIVES TO NASH EQUILIBRIUM

This section briefly discusses and sets aside some alternatives to Nash equilibrium proposed in the literature. Although these alternatives solve some problems with Nash equilibrium, they do not solve the problem we address, namely, the absence of Nash equilibria in certain ideal games. The literature offers many alternatives to Nash equilibrium, but I do not know of any designed to substitute for Nash equilibrium in ideal games without Nash equilibria. The alternatives tackle other problems with Nash equilibrium. Although well motivated, they do not fill the void left by absent Nash equilibria.

Two "refinements" of Nash equilibrium are "perfect" and "proper" equilibrium, as introduced by Selten (1975) and Myerson (1978), respectively. Equilibria of these types are Nash equilibria with special stability properties. These refinements address the selection rather than the absence of Nash equilibria and are motivated by the problem of multiple Nash equilibria. They are proposed for nonideal games in which agents have some chance of making mistakes. The arguments for the refinements have no force in ideal games in which agents are error-free. My arguments against the standard of Nash equilibrium are also arguments against these refinements of Nash equilibrium for ideal games, in which the existence of solutions requires the existence of equilibria. To solve the problem of absent Nash equilibria, we must weaken the standard of Nash equilibrium, not strengthen it, so that equilibria appear where before they were absent.

Another type of refinement of Nash equilibrium is proposed by Selten (1965) for multistage, or extensive-form games. It is called "subgame-perfect equilibrium." A subgame of a multistage game is a game that remains

---

[1] In a nonideal normal-form game, subjective Nash equilibria are likewise strategic equilibria. On the other hand, objective Nash equilibria may fail to be strategic equilibria. But the reasons for the failure, such as ignorance of payoffs, also provide good reasons for disqualifying the objective Nash equilibria as solutions in our subjective sense. So the potential conflict between the standards of strategic equilibrium and objective Nash equilibrium does not threaten the standard of strategic equilibrium.

213

after zero or more stages of the game have been played. A subgame-perfect equilibrium is a strategy profile of a multistage game such that the strategy profiles it yields in subgames are Nash equilibria of the subgames. An immediate consequence of the definition is that every subgame-perfect equilibrium is a Nash equilibrium.

The argument for this refinement of Nash equilibrium is very strong but turns on the special features of multistage games. The refinement does not help with the problem of absent Nash equilibria in single-stage games. It reduces to Nash equilibrium in those games. Moreover, granting the arguments for weakening the standard of Nash equilibrium in single-stage games, there must be an analogous weakening of the standard of subgame-perfect equilibrium for multistage games. This makes the refinement inadequate for our purposes.[2]

Besides refinements of Nash equilibrium, the literature proposes alternative types of equilibrium that take into account the beliefs of agents. Let us consider the alternative equilibria as replacements for missing Nash equilibria in ideal games.

Skyrms's (1990a: 28–32) account of "deliberational dynamics" restricts itself to games in which agents are nonideal and have only "bounded rationality." In this respect it is similar to work of Harsanyi and Selten (1988, Chp. 4) and Binmore (1990, Sec. 6.3) deriving the realization of Nash equilibrium in certain games from the agents' bounded rationality. In Skyrms's model, agents begin deliberation by making initial probability assignments to the strategies of all the agents, including themselves. These assignments are made in isolation from considerations of strategy since the agents' bounded rationality prohibits their dealing with all relevant considerations at once. After the initial probability assignments, agents make revised assignments to take account of strategic considerations. Because of cognitive limitations, the process of revision takes place in stages, with each stage used as input for a rule of bounded rationality that takes the agents to the next stage. In suitable conditions the process of revision converges to a "deliberational equilibrium," where revision ceases. When all the agents in a game reach a deliberational equilibrium, they achieve a "joint deliberational equilibrium." Can joint deliberational equilibria substitute for missing Nash equilibria in ideal games?

The dynamic model of deliberation in terms of which joint deliberational equilibria are defined does not fit ideal games. Rules for revision of deliberational states do not apply in an ideal game because there are no stages of deliberation; choice is instantaneous. Suppose, however, the rules for

---

[2] Other refinements of Nash equilibrium for multistage games with similar deficiencies from our point of view are "sequential equilibrium" and "stable equilibrium" as introduced, respectively, by Kreps and Wilson (1982) and Kohlberg and Mertens (1986).

revision of deliberational states are adjusted for ideal games and interpreted so that a deliberational equilibrium is a choice for which there are no second thoughts. Can joint deliberational equilibria then replace missing Nash equilibria? No, some ideal games lack joint deliberational equilibria under this interpretation. In an ideal version of Matching Pennies in pure strategies every profile includes a choice that prompts second thoughts, assuming that second thoughts follow expected utility increases. Whatever profile is realized, some agent has evidence that he would have gained utility by choosing differently. Hence no profile is a joint deliberational equilibrium. This alternative type of equilibrium does not solve the problem of absent Nash equilibria.

Aumann (1974, 1987) considers games in which agents have unbounded rationality but may have imperfect information about the strategies of other agents. In the earlier article he introduces "correlated equilibrium" for cooperative games, and in the later article he extends this type of equilibrium to noncooperative games. A correlated equilibrium is defined in terms of "correlated strategies." A correlated strategy is a probability mixture of profiles of pure strategies.[3] A correlated equilibrium is a correlated strategy such that no agent has an incentive to deviate unilaterally – each agent's part maximizes expected utility given the other agents' parts. The correlated equilibrium is objective or subjective depending on the type of probabilities involved. We are interested in subjective correlated equilibria as possible replacements for missing subjective Nash equilibria.

Despite its ingenuity, Aumann's work does not resolve the problem we address, the absence of Nash equilibrium in some ideal games. The existence of equilibria in ideal games is not ensured by taking equilibrium to be correlated equilibrium. Correlated equilibrium is weaker than Nash equilibrium but still does not exist in every ideal game. In an ideal version of Matching Pennies in pure strategies agents know the profile realized, so the only correlated strategies are the profiles themselves, represented by the trivial probability distributions that assign some profile probability 1 and other profiles probability 0, or, in other words, assign pure strategies in the profile probability 1 and other pure strategies probability 0. None of these correlated strategies is a correlated equilibrium since given any profile some agent has an incentive for unilateral deviation.

Another sort of equilibrium, introduced by Bernheim (1984) and Pearce (1984), is for noncooperative games in which agents have common

---

[3] A correlated strategy is not a combination of agents' mixed strategies. It is represented by a probability distribution over profiles of pure strategies, not by a profile of, for each agent, a probability distribution over the agent's pure strategies. In a noncooperative game it may arise if each agent responds to observation of the same random variable. It does not require a cooperative framework in which agents may agree on a joint mixed strategy.

knowledge of the payoff matrix and of maximization of expected utility by agents. In a two-person game they call an agent's strategy "rationalizable" if and only if it maximizes expected utility with respect to a probability assignment to strategies of his opponent that has her maximize expected utility with respect to a probability assignment to his strategies that has him maximize expected utility with respect to a probability assignment to her strategies, and so on. This terminology is generalized for n-person games. Also, a profile is said to be "rationalizable" if and only if it comprises only rationalizable strategies. A rationalizable profile is an equilibrium in the sense that if it is realized, then given the idealizations concerning agents each agent's strategy fits his *beliefs* about the strategies of other agents. The profile is a subjective Nash equilibrium since given the profile no agent has an incentive for a unilateral change of strategy.

Every objective Nash equilibrium is a rationalizable profile (Bernheim, 1984: 1018, Pearce, 1984: 1034). However, not every rationalizable profile is an objective Nash equilibrium (Pearce, 1984: 1035). Hence rationalizable profiles are more common than objective Nash equilibria. In fact, in every finite noncooperative game, some profile is rationalizable (Bernheim, 1984: 1015, Pearce, 1984: 1033). Despite this result, rationalizability does not solve the problem of absent Nash equilibria in ideal games.

In ideal games agents have more information about the profile realized than is provided by common knowledge of the payoff matrix and of maximization of expected utility. They are prescient. This strong idealization prevents defective outcomes in which some agent's strategy rests on misinformation about the other agents. Given the agents' information in ideal games, we suppose that each agent accurately attributes probabilities to the other agents' strategies. The probabilities he attributes correctly describe the mixed strategies the other agents adopt. Consequently, we have to disqualify as equilibria rationalizable profiles that involve inaccurate probability assignments. We say that an agent's strategy is "rationalized" if and only if the definition of rationalizability holds with respect to the agents' actual probability assignments. Furthermore, we say that a profile is "rationalized" if and only if it comprises only rationalized strategies. Only the rationalizable profiles that are rationalized may count as equilibria in ideal games.

Once we take an equilibrium to be a rationalized profile, the problem of absent equilibria reappears. In an ideal version of Matching Pennies in pure strategies, all the profiles are rationalizable – the rationalizable profiles are the same as the profiles left after iterative removal of strictly dominated strategies (Pearce, 1984: 1035) – yet none of the rationalizable profiles is rationalized. None is composed exclusively of strategies that maximize expected utility with respect to actual probability assignments. The agents' prescience entails certainty about the profile realized. Hence, the only possible

216

probability assignments have both agents assign probability 1 to some profile and probability 0 to the other profiles. No profile is rationalized with respect to such pairs of probability assignments. The profile (H, T), for instance, is rationalized with respect to probability assignments that have the first agent, the matcher, certain of (H, H) and the second agent, the mismatcher, certain of (H, T). But these probability assignments are not possible in an ideal version of Matching Pennies. They cannot both be accurate since they attribute different probabilities to the second agent's strategies.

The alternative types of equilibria – joint deliberational equilibria, correlated equilibria, and rationalizable profiles – are best suited for nonideal games in which agents make nontrivial probability assignments to the strategies of other agents. Even there, some revisions are needed to take account of the possibility of insufficient incentives to switch strategies.

8.3. STRATEGIC EQUILIBRIUM AND NONDOMINATION

Next, let us compare the standard of strategic equilibrium with the standard of nondomination. The standard of nondomination we consider rejects as a solution any profile with a strategy that is strictly dominated. This formulation has more cogent support than formulations in terms of ordinary dominance. The standard of nondomination provides an especially interesting comparison with the standard of strategic equilibrium since, unlike the latter, it does not draw on the information about the choices of other agents that an agent obtains from his own choice. In the case of a strictly dominated strategy, the standard of nondomination states that this information is superfluous. It says that whatever the information reveals, a strictly dominated strategy is not part of a solution. The standard of nondomination is also simpler technically than the standard of strategic equilibrium. Since the standard of nondomination puts aside information carried by assumptions about choices, it uses nonconditional preferences rather than preferences conditional on choices.

Our comparison of the standards of nondomination and strategic equilibrium uses our search procedure to identify strategic equilibria. So we limit our conclusions to games that meet the search procedure's restrictions, and so have sequences of nearest alternative profiles that follow incentives exclusively.

8.3.1. The Standard of Nondomination

Let us more fully present the standard of nondomination. The standard disqualifies as a solution a profile that assigns to an agent a strategy that pays less than an alternative strategy no matter what other agents do. It requires that a solution not contain a strategy that is strictly dominated, that is, pays less than an alternative strategy in all cases. One strategy strictly dominates another if and only if it pays better in all cases, and a strategy is *not* strictly dominated if and only if for every alternative it pays at least as well as the alternative in some case. The standard of nondomination is an objective standard since it rests on comparisons of payoffs rather than incentives. But in the ideal games we treat, incentives coincide with payoff increases so that it can be taken as a subjective standard as well.

Our standard of nondomination is better supported than the ordinary standard of nondomination. Requiring that a solution lack strictly dominated strategies is less stringent than requiring that it lack strategies dominated in the ordinary sense. A strategy dominated by another in the ordinary sense pays worse in some cases and no better in others. It does not necessarily pay less than the dominant strategy in every case. If the strategy and the dominant strategy pay equally well in cases that are certain to obtain, the two strategies have the same utility for an informed agent and rationality does not require choosing the dominant strategy.

Strategic considerations may even favor a strategy dominated in the ordinary sense. Consider an ideal normal-form game in pure strategies with the payoff matrix in Figure 8.1. Figure 8.2 represents the incentive structure of the payoff matrix using a box and arrow diagram. R2 dominates R1 in the ordinary sense, but R1 is still a good strategy since given it C1 is expected. (R1, C1) is an attractive candidate for a solution. It is a Nash and strategic equilibrium although R1 is dominated. (R2, C2) is also a Nash and strategic equilibrium, but (R1, C1) is more attractive, being Pareto superior to (R2, C2). The standard of nondomination in the ordinary sense disqualifies (R1, C1) as a solution, but the standard of nondomination in the strict sense does not. It disqualifies a profile as a solution only if the profile contains a

2, 2   0, 1

2, 1   1, 2

Figure 8.1   Dominance opposed.

Figure 8.2   Incentives.

218

strictly dominated strategy. Since R2 does not strictly dominate R1, (R1, C1) meets the standard.

The case against the standard of ordinary nondomination is even more powerful in the following example: In it the profile that is the unique Nash equilibrium, the unique strategic equilibrium, and the intuitive solution involves a dominated strategy. Consider an ideal normal-form game in pure strategies with the payoff matrix in Figure 8.3. The box and arrow diagram for the matrix is in Figure 8.4. R2 dominates R1, but (R1, C1) is the unique Nash equilibrium. We assume a complete version of the game where optimal incentives are selected, and in the case of (R3, C1) the incentive to switch to R1 is selected over the incentive to switch to R2. Then there is a terminating group path of incentives away from every profile except (R1, C1). Figure 8.5 brings this out. Every profile except (R1, C1) initiates a group path that terminates in (R1, C1). Take (R1, C2), for instance. It initiates the path (R1, C2), (R3, C2), (R3, C1), (R1, C1). This path terminates

3, 3	1, 2	1, 1
3, 1	2, 2	2, 3
2, 3	3, 2	3, 1

Figure 8.3   Dominance defeated.

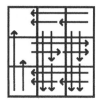

Figure 8.4   Incentives of the payoff matrix.

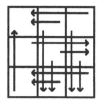

Figure 8.5   Incentives of the complete version.

219

with (R1, C1). Take also (R2, C1). This profile initiates the terminating path (R2, C3), (R3, C3), (R3, C1), (R1, C1). Since every profile except (R1, C1) starts a terminating group path in the complete version, (R1, C1) is the game's unique strategic equilibrium. Intuitively, (R1, C1) is the game's solution even though it fails to meet the standard of ordinary nondomination.

Note that the example obeys the restriction on nearest alternative profiles that governs our search procedure for strategic equilibria. That restriction allows nearest alternative profiles to follow one of two tying incentives, as in the selection of the incentive from (R3, C1) to (R1, C1) rather than to (R2, C1). It does not allow movement from one tying option to another, as in a move from (R2, C1) to (R1, C1). In conformity with the restriction, the example takes the nearest alternative profile for (R2, C1) to be (R2, C3), not (R1, C1).

We can also verify by first principles that (R1, C1) is the unique strategic equilibrium. The agents' completely reduced incentive structures confirm its status. Take the path for Row started by (R1, C2). It is (R1, C2), [(R3, C2)] (R3, C1), (R1, C1). This path terminates in the last profile. All agent paths initiated by profiles other than (R1, C1) terminate in similar fashion. Only (R1, C1) is a strategic equilibrium since only that profile does not initiate a terminating path in the agents' completely reduced incentive structures.

The foregoing example makes a strong case against the standard of nondomination in the ordinary sense. We replace it with the standard of nondomination in the strict sense and compare the standard of strategic equilibrium only with the standard of nondomination in the strict sense. Henceforth we take it as understood that when, for brevity, we speak of domination and nondomination we mean strict domination and its absence.

The standard of nondomination is inspired by the decision principle of dominance, or nondomination. The decision principle, discussed in Section 4.5, states that given a finite number of contending options and taking causal independence of options and states to be presumed by strict domination, a strictly dominated option is irrational. This decision principle requiring nondomination concerns preferences and directly yields a subjective standard of nondomination for games. It also yields an objective standard of nondomination for ideal games, where agents' rationality and knowledge of payoffs generate preferences agreeing with payoff comparisons. Since the decision principle of nondomination is restricted to cases in which options do not causally influence states, the standard of nondomination for games is also restricted to cases in which an agent's strategies do not causally influence other agents' strategies. This restriction is met in normal-form games. Moreover, since the decision principle of nondomination is

$$1, 1 \qquad 1, 0$$

$$2, 1 \qquad 2, 0$$

$$3, 1 \qquad 3, 0$$

$$\cdot \qquad \cdot$$

$$\cdot \qquad \cdot$$

$$\cdot \qquad \cdot$$

Figure 8.6   Incentives without end.

restricted to cases in which an agent has a finite number of options, the standard of nondomination for games is restricted to cases in which agents have a finite number of strategies. To ensure satisfaction of assumptions in the argument for the standard of nondomination, our comparison of the standard of strategic equilibrium with the standard of nondomination is limited to ideal, finite, normal-form games.

To illustrate the need for restricting the standard of nondomination to games in which agents have a finite number of strategies, take an ideal normal-form game with the payoff matrix in Figure 8.6. Each of Row's strategies is dominated by all those below it. None of them is undominated. Hence no profile meets the standard of nondomination. In a game in which an agent has an infinite number of strategies, every profile may have a strategy that is strictly dominated. Given that every ideal game has a solution, the example shows that meeting the standard of nondomination is not in general a necessary condition for being a solution. In this game a solution includes a strictly dominated strategy.[4]

---

[4] Let us find the strategic equilibria in this example. There is a terminating path for Column from C2 to C1, but no terminating path from C1 to C2, whatever Row's response. C1 is Column's part in any strategic equilibrium. Now consider Row's strategies. Every dominated strategy starts a path to a strategy that dominates it. But the path from one of Row's dominated strategies to a dominant strategy does not terminate, since the dominant strategy is dominated by a strategy that is dominated, and so on, ad infinitum. None of Row's strategies initiates a terminating path of incentives in the incentive structure of the payoff matrix or any partial version derived from it. Each initiates a path of incentives for Row that does not end, and none of those paths terminates in a tree, since no closed tree meets the minimality condition. Each closed tree has a member that does not appear in the closed subtree it immediately precedes. For instance, in a partial version in which the incentives pursued produce unit increases, the path away from R1 is the following: R1, R2, R3, ... This path does not end and does not terminate in a subpath. No closed subpath includes its immediate predecessor. In the complete version, in which pursuit of incentives is not relentless, the profiles where pursuit stops are strategic equilibria. They meet the standard of strategic equilibrium but fail to meet the standard of nondomination.

$$2, 2 \quad 0, 1$$

$$1, 2 \quad 1, 1$$

Figure 8.7   Nondomination without strategic equilibrium.

8.3.2. *Logical Relations*

To begin comparing the two standards for solutions, let us consider whether meeting the standard of nondomination is sufficient for meeting the standard of strategic equilibrium. Can a profile meet the standard of nondomination but fail to meet the equilibrium standard? Can a profile without strictly dominated strategies start a terminating path for some agent? The answer is, yes. A strategy in a profile initiates a path for an agent if given the profile the agent prefers another strategy. The agent need not prefer the other strategy no matter what other agents do. If his path of incentives terminates, the profile is not a strategic equilibrium.

An ideal normal-form game with the payoff matrix in Figure 8.7 provides an illustration. The profile (R2, C1) meets the standard of nondomination but fails to meet the standard of strategic equilibrium. Neither R2 nor C1 is strictly dominated, but R2 does start a terminating path in every complete version of the payoff matrix. To see this, note that C1 strictly dominates C2. Strict dominance may not be decisive in every case, but it settles Column's choice here. Whatever Row does, Column's response is C1. Row then has an incentive to switch from R2 to R1 and no incentive to switch again. This case shows that nondomination is insufficient for strategic equilibrium.

Next, let us consider whether meeting the standard of nondomination is a necessary condition for meeting the standard of strategic equilibrium. Can a profile meet the standard of strategic equilibrium but fail to meet the standard of nondomination? Can a profile fail to initiate a terminating path for any agent, yet include a strategy that is strictly dominated? Let us begin by considering whether nondomination is necessary for strategic equilibrium *if agents pursue only optimal incentives*. We can show that it is necessary in this case given our standing assumption that the games treated are ideal, finite, and normal-form and have nearest alternative profiles that follow incentives exclusively.

Take a game meeting our assumptions and consider a profile with a strictly dominated strategy. We want to show that it is not a strategic equilibrium. First, suppose that in the partial version of the game the agent's incentive to switch to his strictly dominant strategy starts a group path. The group path never returns to the profile. For no matter how other agents respond to the agent's strategy, he never switches back to the dominated strategy. He never

222

switches back since by assumption he pursues optimal incentives, and however other agents respond to his strategies, his payoff is larger if he switches to the strictly dominant strategy (or stays with it) rather than if he switches to the strictly dominated strategy. Since the group path does not return to its initial profile, it must terminate. If the path does not terminate in a profile, it is endless. And if it is endless, it must cycle, since a finite game has only a finite number of profiles and a finite number of ways in which they can be arranged. Since the cycle does not include the path's initial profile, the cycle has an immediate predecessor not in the cycle. Hence the path terminates in the cycle.

Next, suppose that in the partial version the agent's incentive to switch to his strictly dominant strategy does not start a group path. Then that incentive is preempted by another incentive that obtains with respect to the profile. The interagent selection rule excludes the possibility that the agent's incentive is preempted by an insufficient group incentive, for replacing the selected incentive with the agent's incentive to switch to his strictly dominant strategy makes the profile start a terminating group path, whatever the structure of the partial version, as the previous paragraph shows. So the selected incentive must start a terminating group path. Hence no matter whether the agent's incentive starts a group path or not, the profile starts a terminating group path in the game's partial version.

By the interagent stopping rule, the initial incentive of a terminating group path away from a profile in the partial version is included in the complete version. Hence in the complete version there is a group path, and therefore a terminating group path, away from the profile containing the strictly dominated strategy. According to our search procedure, the profile is not an equilibrium of the game. Therefore no profile with a strictly dominated strategy is a strategic equilibrium in an ideal normal-form game meeting our assumptions. Strategic equilibria exclude dominated strategies. A profile must meet the standard of nondomination to meet the standard of strategic equilibrium given pursuit of optimal incentives.

Now let us put aside the assumption that agents pursue optimal incentives. Given the possibility of pursuit of nonoptimal incentives, nondomination is *not* necessary for strategic equilibrium. To see this, take an ideal normal-form game with the payoff matrix in Figure 8.8, a repetition of Figure 4.9. Suppose that, assuming relentlessness, agents pursue optimal incentives except that given C2 Row pursues his incentive to switch from R3 to R1 instead of his optimal incentive to switch to R2. The game's partial version then has a path going counterclockwise around the four corners of the payoff matrix. See Figure 8.9. The partial version satisfies the interagent selection rule. The group path started by (R1, C1) cycles back to that profile, so Row's incentive to switch from R1 to R3 given C1 is an insufficient group incentive. But so is Row's alternative incentive to switch from R1 to R2 if substituted

$$1, 2 \quad 1, 1$$

$$2, 2 \quad 2, 1$$

$$3, 1 \quad 0, 2$$

Figure 8.8   Nondomination versus strategic equilibrium.

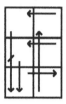

Figure 8.9   Pursuit of nonoptimal incentives.

in the partial version.[5] Similarly, Row's incentive to switch from R3 to R1 given C2 is an insufficient group incentive. But so is his alternative incentive to switch from R3 to R2 if substituted in the partial version.

Next, suppose that in the game's complete version Row's pursuit of incentives stops at (R1, C1), and this is the only profile at which pursuit of incentives stops. See the crossed out arrow in Figure 8.9. The complete version then conforms with the interagent stopping rule since Row's selected but resisted incentive is insufficient in the partial version. Since (R1, C1) is a group equilibrium of the complete version, according to our search procedure it is a strategic equilibrium of the game itself. Column has no incentive to switch from (R1, C1), and Row's incentives are unpursued, discarded by his completely reduced incentive structure, and so do not start terminating paths in that structure. Yet (R1, C1) contains a dominated strategy, R1. Nondomination is therefore not necessary for strategic equilibrium.

8.3.3. Conflict

Our main question is whether the standard of nondomination conflicts with the standard of strategic equilibrium. That is, do games exist where the sets of profiles satisfying each standard are nonempty but do not overlap? Perhaps

---

[5] If substituted, the incentive generates a group path that has two of Row's incentives in succession: his incentive to switch from R1 to R2 followed by his incentive to switch from R2 to R3. Such a group path may violate a rule for jointly rational pursuit of incentives. This would provide a reason for favoring the incentive our partial version incorporates. We leave the issue open. However it is settled we obtain a partial version that, as explained later, creates a challenge to the standard of strategic equilibrium, one the standard meets.

224

there are games in which *every* strategic equilibrium contains a strictly dominated strategy. Then it is impossible to satisfy both standards. Such conflict in fact arises in the game of Figure 8.9. (R1, C1) is the unique strategic equilibrium. All other profiles start terminating group paths in the complete version of the game. But (R1, C1) contains the dominated strategy, R1. Hence the standards of strategic equilibrium and nondomination conflict.

How should we respond to the conflict? One possibility is to dodge the issue by maintaining that in ideal games, where agents are rational, agents pursue only optimal incentives. Pursuit of optimal incentives does indeed have some appeal as a requirement of rationality. But we want to allow for the possibility that strategic considerations justify pursuit of nonoptimal incentives. Moreover, incentives, not just optimal incentives, are the grounds of the standard of nondomination. So the standard does not lose its motivation if agents pursue nonoptimal incentives. Hence even if the two standards do not conflict in ideal games because agents pursue optimal incentives, they conflict in games that are ideal except for pursuit of nonoptimal incentives. How should the conflict be resolved there? A full theory of rationality in games needs an answer to this question.

For the foregoing reasons, we take the conflict between the two standards seriously. We want to know whether the standard of nondomination or strategic equilibrium has the upper hand when they conflict in games that are ideal except perhaps for pursuit of nonoptimal incentives. Which standard should give way?

I stand by the standard of strategic equilibrium and reject the standard of nondomination. The standard of nondomination is just the standard of incentive-proofness applied to cases in which incentives to switch are ensured by dominance. The argument against the standard of incentive-proofness, or Nash equilibrium in ideal games, also undermines the standard of nondomination. Given prescience and agents' certitude about other agents' strategies, the force of dominance reduces to the presence of an incentive to switch. The only reason to comply with the standard of non-domination is that in a profile with a dominated strategy some agent has an incentive to switch. An incentive to switch to a dominant strategy has no more force than any other incentive to switch. The standard of non-domination may have an epistemic advantage over the standard of strategic equilibrium since it can be applied without information about other agents' strategies. But in ideal games, where agents are prescient, this advantage is nullified. In any case, the epistemic advantage does not give incentives to switch to dominant strategies additional force. Conflict with the standard of incentive-proofness arising from conflict with the standard of nondomination is therefore an insufficient reason to reject the standard of strategic equilibrium. The standard of incentive-proofness is not a necessary condition of

225

rational choice, not when pursuit of incentives is futile. This is Chapter 4's main point. Since incentives to switch are not sufficient reasons to switch when their pursuit is futile, incentives to switch to dominant strategies are not sufficient reasons to switch when their pursuit is futile. Because pursuit of incentives holding with respect to a strategic equilibrium is futile, in cases of conflict between the standards of strategic equilibrium and nondomination, pursuit of incentives to switch to dominant strategies is futile, as illustrated by the game of Figure 8.9. Reflections on futility favor the standard of strategic equilibrium wherever it conflicts with the standard of nondomination.

For further confirmation of our rejection of the standard of nondomination, examine the decision principle behind that standard, namely, the decision principle of dominance, or nondomination. Section 4.5 rejected that decision principle. The case against it further erodes support for the standard of nondomination for solutions.

The decision principle of nondomination is supported by the principle to maximize expected utility given an assignment of probabilities to states of the world and otherwise is supported by a generalization of the expected utility principle designed to function in the absence of probability and utility assignments. As Section 4.5 showed, the principle to maximize expected utility needs a ratificationist interpretation to accommodate the information a choice is anticipated to provide. Such an interpretation computes expected utility under the assumption that an option is adopted. The decision principle of nondomination arises from a ratificationist version of the generalized expected utility principle. The derivation simplifies out the assumption that an option is adopted. Since an option that strictly dominates another has higher utility given each state of the world, information about the state of the world carried by assumption of one of the options does not affect their expected utility comparison. The generalized expected utility principle entails the decision principle to reject an option that has a lower expected utility than another under every quantization of belief and desire, and this principle entails the decision principle of nondomination. Thus the decision principle of nondomination arises from an appropriately restricted decision principle of nonconditional maximization of expected utility. But probing deeper, it also arises from a principle of conditional maximization of expected utility under assumptions that make conditionalization superfluous.

Section 4.5's study of the principle to maximize expected utility shows that maximizing expected utility in a ratificationist way is not a necessary condition for rational choice when pursuit of incentives is futile. In such cases an agent does not have to pick an option that maximizes expected utility on the assumption that it is adopted. Self-support with respect to incentives requires only choosing an option that does not start a terminating path of

226

incentives. These points carry over to the derivative decision principle of nondomination. Compliance with the decision principle of nondomination is not a necessary condition for rational choice when pursuit of incentives is futile. The decision principle to reject an option that has a lower expected utility than another under every quantization of belief and desire has exceptions when pursuit of incentives is futile. In cases in which every option is such that if adopted it has a lower expected utility than another option under every quantization of belief and desire, the principle must be violated, and some option that violates it is rational. Cases of this type arise where the number of options is infinite or where the assumption that an option is adopted changes the preference ranking of options. The incentive to switch to a dominant option is sufficient only if pursuit of the incentive is not futile. In cases of conflict with the standard of strategic equilibrium, the standard of nondomination calls for futile pursuit of incentives and receives no support from circumspect decision principles.

One acceptable revision of the decision principle of nondomination considers an agent's completely reduced incentive structure. It states that a rational choice does not initiate a terminating path of incentives to switch to strictly dominant options. The revised principle prohibits options that start one type of terminating path of incentives and so follows from the principle of self-support. It suggests a revised standard of nondomination for solutions to games. The revised standard says that a solution does not initiate for any agent a terminating path of incentives to switch to strictly dominant strategies. An agent's closed path of incentives to switch to strictly dominant strategies ends in a finite game since strict dominance is transitive and inimical to cycles. Furthermore, the closed path terminates in its last profile just in case all incentives to switch from the last profile are absent, not merely incentives to switch to strictly dominant strategies. Thus every terminating path of incentives to switch to strictly dominant strategies is a terminating path of incentives. Consequently, the standard of strategic equilibrium supports the revised standard of nondomination. Any profile ruled out by the revised standard of nondomination is also ruled out by the standard of strategic equilibrium.

The standard of strategic equilibrium fares well in comparisons with familiar standards for solutions to games. It does not conflict with the standard of Nash equilibrium in ideal normal-form games. It is just a weakening of the standard of Nash equilibrium. Also, it does not conflict with the standard of nondomination in ideal normal-form games, except in special cases in which pursuit of incentives is nonoptimal. In those cases the standard of strategic equilibrium has the upper hand. Intuition and argument favor it over the standard of nondomination.

References

Aumann, R. 1974. "Subjectivity and Correlation in Randomized Strategies." *Journal of Mathematical Economics* 1: 67–96.

———. 1987. "Correlated Equilibrium as an Expression of Bayesian Rationality." *Econometrica* 55: 1–18.

Aumann, R., and A. Brandenburger. 1991. "Epistemic Conditions for Nash Equilibrium." Working Paper No. 91-042. Boston: Harvard Business School.

Aumann, R., and M. Maschler. 1964. "The Bargaining Set for Cooperative Games." In M. Dresher, L. S. Shapley, and A. W. Tucker, eds., *Advances in Game Theory, Annals of Mathematics Studies*, No. 52: 443–76. Princeton, NJ: Princeton University Press.

Bernheim, B. D. 1984. "Rationalizable Strategic Behavior." *Econometrica* 52: 1007–28.

Bernheim, B. D., B. Peleg, and M. D. Whinston. 1987. "Coalition-Proof Nash Equilibria. I. Concepts." *Journal of Economic Theory* 42: 1–12.

Binmore, K. 1990. *Essays on the Foundations of Game Theory*. Oxford: Basil Blackwell.

Bratman, M. 1987. *Intention, Plans, and Practical Reason*. Cambridge, MA: Harvard University Press.

Eells, E., and W. Harper. 1991. "Ratifiability, Game Theory, and the Principle of Independence of Irrelevant Alternatives." *Australasian Journal of Philosophy* 69: 1–19.

Gibbard, A. 1992. "Weakly Self-Ratifying Strategies: Comments on McClennen." *Philosophical Studies* 65: 217–25.

Gibbard, A., and W. Harper. 1978. "Counterfactuals and Two Kinds of Expected Utility." In C. Hooker, J. Leach, and E. McClennen, eds., *Foundations and Applications of Decision Theory*, Vol. 1: 125–62. Dordrecht: Reidel.

Harper, W. 1986. "Mixed Strategies and Ratifiability in Causal Decision Theory." *Erkenntnis* 24: 25–36.

———. 1988. "Causal Decision Theory and Game Theory." In W. Harper and B. Skyrms, eds., *Causation in Decision, Belief Change, and Statistics*, Vol. II: 25–48. Dordrecht: Kluwer.

———. 1991. "Ratifiability and Refinements." In M. Bacharach and S. Hurley, eds., *Foundations of Decision Theory: Issues and Advances*, pp. 263–93. Oxford: Basil Blackwell.

Harsanyi, J. 1973. "Games with Randomly Distributed Payoffs: A New Rationale for Mixed-Strategy Equilibrium Points." *International Journal of Game Theory* 2: 1–23.

229

Harsanyi, J., and R. Selten. 1988. *A General Theory of Equilibrium Selection in Games*. Cambridge, MA: MIT Press.

Jeffrey, R. 1983. *The Logic of Decision*, 2nd edition. Chicago: Chicago University Press.

Kadane, J., and P. Larkey. 1982. "Subjective Probability and the Theory of Games." *Management Science* 28: 113–20.

Kohlberg, E., and J. Mertens. 1986. "On the Strategic Stability of Equilibria." *Econometrica* 54: 1003–37.

Kreps, D., and R. Wilson. 1982. "Sequential Equilibria." *Econometrica* 50: 1447–74.

Luce, R. D., and H. Raiffa. 1957. *Games and Decisions*. New York: Wiley.

Myerson, R. 1978. "Refinements of the Nash Equilibrium Concept." *International Journal of Game Theory* 7: 73–80.

Nash, J. 1950. "Equilibrium Points in *N*-Person Games." *Proceedings of the National Academy of Sciences* 36: 48–9.

Nozick, R. 1969. "Newcomb's Problem and Two Principles of Choice." In N. Rescher, ed., *Essays in Honor of C. G. Hempel*, pp. 114–46. Dordrecht: Reidel.

Pearce, D. 1984. "Rationalizable Strategic Behavior and the Problem of Perfection." *Econometrica* 52: 1029–50.

Rabinowicz, W. 1989. "Stable and Retrievable Options." *Philosophy of Science* 56: 624–41.

Resnik, M. 1987. *Choices*. Minneapolis: University of Minnesota Press.

Richter, R. 1984. "Rationality Revisited." *The Australasian Journal of Philosophy* 62: 392–403.

Selten, R. 1965. "Spieltheoretische Behandlung eines Oligopolmodells mit Nachfrageträgheit." *Zeitschrift für die gesamte Staatswissenschaft* 12: 301–24.

1975. "Reexamination of the Perfectness Concept of Equilibrium in Extensive Games." *International Journal of Game Theory* 4: 25–55.

1988. *Models of Strategic Rationality*. Dordrecht: Kluwer.

Sen, A. 1970. *Collective Choice and Social Welfare*. San Francisco: Holden-Day.

Shin, H. 1991. "Two Notions of Ratifiability and Equilibrium in Games." In M. Bacharach and S. Hurley, eds., *Foundations of Decision Theory*, pp. 242–62. Oxford: Basil Blackwell.

Skyrms, B. 1990a. *The Dynamics of Rational Deliberation*. Cambridge, MA: Harvard University Press.

1990b. "Ratifiability and the Logic of Decision." In P. French, T. Uehling, and H. Wettstein, eds., *The Philosophy of the Human Sciences, Midwest Studies in Philosophy*, Vol. 15: 44–56. Notre Dame, IN: University of Notre Dame Press.

Sobel, J. H. 1976. "Utilitarianism and Past and Future Mistakes." *Noûs* 10: 195–219.

1982. "Utilitarian Principles for Imperfect Agents." *Theoria* 48: 113–26.

1990. "Maximization, Stability of Decision, and Actions in Accordance with Reason." *Philosophy of Science* 57: 60–77.

Sobel, J. H. 1994. *Taking Chances: Essays on Rational Choice*. Cambridge: Cambridge University Press.

Stalnaker, R. 1994. "On the Evaluation of Solution Concepts." *Theory and Decision* 37: 49–73.

230

Vallentyne, P. 1991. "The Problem of Unauthorized Welfare." *Noûs* 25: 295–321.

Von Neumann, J., and O. Morgenstern. 1944. *Theory of Games and Economic Behavior*. Princeton, NJ: Princeton University Press.

Weirich, P. 1985. "Decision Instability." *The Australasian Journal of Philosophy* 63: 465–72.

   1988. "Hierarchical Maximization of Two Kinds of Expected Utility." *Philosophy of Science* 55: 560–82.

   1990. "L'Utilité Collective" (Collective Utility). In E. Archambault and O. Arkhipoff, eds., *La Comptabilité Nationale Face au Défi International*, pp. 411–22. Paris: Economica.

   1991a. "Group Decisions and Decisions for a Group." In A. Chikán, ed., *Progress in Decision, Utility and Risk Theory*, pp. 271–9. Norwell, MA: Kluwer.

   1991b. "The General Welfare as a Constitutional Goal." In D. Speak and C. Peden, eds., *The American Constitutional Experiment*, pp. 411–32. Lewiston, NY: Edwin Mellen Press.

   1994. "The Hypothesis of Nash Equilibrium and Its Bayesian Justification." In D. Prawitz and D. Westerståhl, eds., *Logic and Philosophy of Science in Uppsala*, pp. 245–64. Dordrecht: Kluwer.

231

Index

Entries for technical terms indicate pages where the terms are explained.

233

Jeffrey, R., xii, 56, 58, 72

Kadane, J., 71–2
Kohlberg, E., 214
Kreps, D., 214

Larkey, P., 71–2
Luce, R., xii, 118, 207

Maschler, M., ix
Matching Pennies, 2, 13–14, 41–2, 61–4, 70–1, 74–5, 110–11, 150–3; complete version of, 181, 186, 192–6; embedded, 87–8; mixed extension of, 76, 123–4; partial version of, 169–72
maximization of expected utility, 115–16, 123–4
Mertens, J., 214
Morgenstern, O., 37, 58
Myerson, R., 213

Nash equilibrium, 3; multiple Nash equilibria, 75, 149–50, 203–4; nonexistence of, 62; objective, 3, 61; subjective, 3, 61, 78; unattractive, 66, 198–202
Nash, J., 4
Newcomb-like phenomena, 85
Newcomb's problem, 58, 85, 114
Nozick, R., 58

Pareto optimality, 75
Pareto superiority, 75
path of incentives, 85; for an agent in a complete version, 185; closure of, 85, 180; for a group, 161; for a group and compressed, 184–7; with indecisive switching, 113, 138, 162, 224; subpath of, 86
payoff, 7, 11
payoff matrix, 9, 11; ignorance of, 202–3
Pearce, D., 215–16
principle alpha, 117–18, 127–9, 209–10
principle of rejection, 81–4; see also principle of sufficiency
principle of sufficiency, 79; basis, 81, 82, 91; recursive, 81, 83; see also principle of rejection; rationality, necessary condition of
profile of strategies, 7; nearest alternative to, 173; remainder of, 134; response, 137; switch, 137; terminal, 155
pursuit of incentives, 95–6, 169, 173;

cyclical, 64, 177; relentless, 166–7, 169–71, 175

quantization, 115

Rabinowicz, W., 56, 125
Raiffa, H., xii, 118, 207
ratification, 56; principle of, 56–9, 116, 122–3
rationality, 31; bounded, 214; compensatory standards of, 25–6; conditional, 24–8, 75; full, 31, 70, 101–2; joint, 20–3; necessary condition of, 79–80, 85–7, 93, 97, 103, see also principle of sufficiency; noncompensatory standards of, 25–6, robustness of, 31–2; see also idealization of rationality
representation of a game, 14; appropriate abstract, 16–17; complete, 178–9; cooperative, 9; noncooperative, 9; partial, 166
resisters, 193
Resnik, M., xii
response: nonrelative, 134; relative, 133
Richter, R., 72

search procedure, 159–60
selection rule, 107
self-defeat, 54, 60, 74–5, 104
self-support, 5, 53, 61, 68; joint, 48, 51; nonrelative, 48; principle of, 104; relative, 48, 154; satisfaction of the principle of, 105–6; strategic, 104, 130
Selten, R., 9, 131, 213, 214
Sen, A., 117
Shin, H., 58
Skyrms, B., 58, 214
Sobel, J., 31, 56, 100, 205
solutions of a game, 23; consistency of, 16–17; cooperative and noncooperative standards for, 17; as intensional objects, 18; nonrelative, 23; objective, 19–20, 29–30; potential, 15; relative, 23; subjective, 19–20, 30
Stalnaker, R., 15, 32
stopping rule, 109
strategic equilibrium, 130, 144; for agents, 161, 163; for a group, 161, 163; nonrelative, 49, 136; relative, 49, 136
strategic reasoning, 7–8, 38, 39–40, 51–2
strategy, 6; mixed, 4; pure, 4; terminal, 154

234